ASBESTOS IN THE NATURAL ENVIRONMENT

Other volumes in this series

Studies in Environmental Science 37

ASBESTOS IN THE NATURAL ENVIRONMENT

H. Schreier

Department of Soil Science and Westwater Research Centre, University of British Columbia, Vancouver, B.C. V6T 2A2, Canada

ELSEVIER
AMSTERDAM — OXFORD — NEW YORK — TOKYO   1989

ELSEVIER SCIENCE PUBLISHERS B.V.
Sara Burgerhartstraat 25
P.O. Box 211, 1000 AE Amsterdam, The Netherlands

Distributors for the United States and Canada:

ELSEVIER SCIENCE PUBLISHING COMPANY INC.
655, Avenue of the Americas
New York, NY 10010, U.S.A.

ISBN 0-444-88031-3
ISBN 0-444-41696-X (Series)

CONTENTS

Pages

LIST OF FIGURES:

LIST OF TABLES:

PREFACE

Asbestos minerals have novel properties that make them highly desirable for industrial use yet hazardous to human health and undesirable for plant growth. The combination of heat resistance, non-combustibility, unusual tensile strength and resistance to selective chemical attack have favoured asbestos to be used in virtually thousands of industrial applications. In its various forms, it is an ubiquitous substance and occurs as a natural fiber with all the advantages pertaining thereto; it can be woven, molded and added to other materials to form superior and special purpose products, whose value, since the advent of the industrial society, has been greatly enhanced. But, inherent in the very uniqueness, are those obvious disadvantages and hazards well documented in the medical and general health literature. Also less publicized, asbestos materials have significant deleterious effects on soil and plant ecology, giving rise to very precarious and novel ecosystems.

To understand the environmental problems associated with asbestos fibers medical researchers have focused their attention on its physical properties and now accept that fiber size, geometry, type, durability and dose are of primary concern. The medical literature which is mostly dedicated to occupational exposure to asbestos is massive, and the hazards associated with airborne exposure are also well documented. Considerably less attention has been devoted to non-occupational exposure and environmental effects relating to animal and plant growth in the natural environment. Soil and plant ecologists have concentrated some efforts upon understanding the chemical characteristics of asbestos rich material and their effect on plant growth. Major nutrient imbalances and excess concentrations of trace metals have been identified as main causes for the poor plant response.

The medical and ecology researchers have taken different approaches to study the effects of asbestos on man, animals, and plants. In spite of the enormous research efforts a comprehensive understanding of the human, animal and plant health problems has so far eluded us. Ecological and environmental research involving asbestos fibers has only started in the past 25 years. Given the complexity of the asbestos analysis and the extent of the problem an attempt is made in this book to bring together the multitude of subjects pertaining to asbestos in the natural environment.

Many people have contributed to this book and their assistance is gratefully acknowledged. James Taylor stirred my interest in the topic and gave valuable advice in the preparation of the manuscript. Patricia acted as a most patient and diligent editor, Sandra Brown produced the figures and assisted in the production of the manuscript and Genevieve kept a smiling face when bedtime stories had to be shortened. Much reference material was provided by Mme. S. Sokov, Informatheque, Programme de Recherche sur l'amiante, Universite de Sherbrooke (PRAUS). Finally a special thank you goes to Les Lavkulich for giving me the freedom to delve into the many different fields of asbestos research for which I was inadequately prepared.

Vancouver, British Columbia, Canada   H. Schreier
January, 1989

CHAPTER 1

GENERAL INTRODUCTION

1.1 <u>Introduction</u>

Investigating asbestos materials in the environment is a truly interdisciplinary task. The fibers are of very small dimensions and find their way into all components of the environment (see Figure 1). Progress in asbestos research has been slow because a quantitative analysis of asbestos fibers is tedious, time consuming and costly. In addition, the mineralogy of asbestos is complicated because fibers generally develop as a result of defects in the mineral structure, they are subject to many isomorphic substitutions, and most fibers contain impurities. Brucite, magnetite and carbonates are often intergrown into the mineral structure of asbestos fibers. To investigate environmental aspects of asbestos therefore requires input from many specialized disciplines. Medical researchers have investigated asbestos related problems by epidemiological studies and in Vivo and in Vitro experiments. The physio-chemical properties of asbestos have generally been investigated by mineralogists, chemists, and microscopists, while plant physiologists have focused their attention on the effects of asbestos on species distribution and plant vigour. Ecological work has to draw on all these specialists in order to arrive at a comprehensive understanding of the effects of asbestos on the environment.

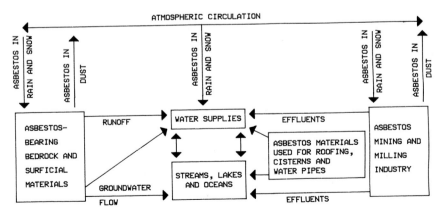

Fig.1. Flowchart of asbestos fiber pathways in the environment

The novel physical and chemical properties of asbestos have made it a very desirable and useful industrial material. The analytical difficulties in quantifying asbestos, the health hazards associated with airborne exposure, and the adverse effects on plant growth make environmental assessments of asbestos both challenging and fascinating.

1.2 What is Asbestos ?

More than 30 mineral silicates can crystallize into fibrous forms but only a few of these are considered asbestos minerals. Asbestos can be defined as cotton like fibrous minerals which have unusual tensile strength, are heat resistant, and have length to width ratios that are greater than 3:1. Most asbestos fibers have mineral equivalents that are massive and non-fibrous and simple chemical analysis is not sufficient to characterize asbestos minerals. A list of the asbestiform minerals and their non-fibrous equivalents is provided in Table 1.

TABLE 1

Fibrous and non-fibrous forms of asbestos minerals

Mineral Type	Fibrous Form	Non-Fibrous Form	General Chemical Formula
Serpentine	Chrysotile	Antigotite Lizardite	$Mg_6[(OH)_4Si_2O_5]_2$
Amphibole	Actinolite	Massive-Actinolite	$Ca_2Fe_5(OH)2Si_8)_{22}$
	Anthopholite	Massive-Anthopholite	$(Mg,Fe)_7[(OH)Si_4O_{11}]_2$
	Amosite	Grunerite	$Fe_7(OH)_2Si_8O_{22}$
		Commingtonite	$Mg_7(OH)_2Si_8O_{22}$
	Tremolite	Massive-Tremolite	$Ca_2Mg_5(OH)_2Si_8O_{22}$
	Crocidolite	Riebeckite	$Na_2Fe_5[(OH)Si_4O_{11}]_2$

The term asbestos is used commercially and relates to all types of asbestos. However, there are considerable mineralogical, morphological and physio-chemical variances among the different types of fibers. Based on mineralogy asbestos can best be divided into serpentine and amphibole minerals and these will be discussed separately.

1.2.1 Serpentine Asbestos

Serpentine minerals can be divided into massive and acicular minerals and the latter form is known as chrysotile. Chrysotile is the most widely distributed asbestos mineral. It has a layered structure made up of SiO_4 tetrahedral and $Mg(OH)_2$ octahedral layers. The crystallographic mismatch between these two types of layers is responsible for a curvature in the structure which results in the characteristic cylindrical or tubular form of the chrysotile fibers. It is now generally agreed that chrysotile is less likely to occur in serpentine materials with high aluminum content. Aluminum substitution for magnesium and silica generally results in the flatter layered non-fibrous serpentine minerals antigorite and lizardite. Serpentine minerals are closely associated with ultramafic rocks particularly where dunite (dominantly olivine) and peridotite rock (dominantly pyroxene and olivine) are metamorphosed and hydrothermally altered to form chrysotile, antigorite, lizardite, brucite and magnetite minerals. As shown by Ross (1982), Harben and Bates (1984) and Brooks (1987) the distribution of serpentine minerals is quite widespread but the acicular quality for commercial use is limited to a few deposits.

1.2.2 Amphibole Asbestos

Amphibole minerals have often been referred to as the waste baskets of mineralogy because of the many isomorphic substitutions that take place in the formation of these minerals. Amphiboles are common in the dark coloured igneous and metamorphic rocks and the elemental substitutions lead to many structural faults and crystallographic disorders. Most amphibole minerals originate from metamorphic processes in which pyroxene minerals are hydrated to anthopholite and tremolite. Metamorphic alterations of sedimentary silicates and iron formations have also been found to produce amphiboles. Attempts to explain fibrosis have been made by measuring surface energy as determined by the number and strength of broken bonds per unit area. More convincing results were obtained from aluminum determinations which, as in the case of fibrous serpentine, were negatively correlated with amphibole fibrosity. Few aluminum substitutions result in high fibrosity while large aluminum substitutions produce massive amphibole minerals. In contrast to chrysotile, amphibole fibers are solid rods and they are often stiffer than their chrysotile counterparts.

Amphiboles are generally distinguished on the basis of some idealized end member such as: Iron-magnesium amphibole (grunerite and amosite), calcium amphiboles (tremolite) and alkaline amphibole (riebeckite and crocidolite).

In contrast to chrysotile, amphibole fibers are acid resistant but the resistivity varies with amosite being least acid resistant to crocidolite and tremolite which can tolerate strongly acidic environments (Hodgson 1986).

1.3 A brief history of asbestos use

There is evidence that asbestos materials have been used since prehistoric times. Lee and Selikoff (1979), in their historical review of asbestos, indicated that the Greek meaning of asbestos signifies "unquenchable" or "indestructible" and the earliest record of asbestos use by man is attributed to Finland where asbestos based pottery was supposedly in use some 4500 years ago. Asbestos based pottery was also reported to be used in Sudan and Kenya before the middle of the first century B.C. and records for the same time period revealed that the Greeks used asbestos as lamp wicks. The Egyptians used asbestos as funeral dresses for kings and the Romans used asbestos for cremation. The story of Charlemagne impressing his guests after a banquet by throwing the asbestos table cloth into the fire to clean it has been quoted by many authors. Asbestos based cloth was cited by Lee and Selikoff (1979) and the Committee on Non-Occupational Health (1984) to be in use in Norway, the USSR, China, and Italy before 1860.

With the advent of steam energy in the late 1860's the market for asbestos significantly changed and by 1890 the manufacturing of asbestos products was in full swing. Mines in Canada and South Africa, Russia, USA, Italy and Rhodesia were the primary sources and by the late 1940's asbestos was used in virtually hundreds of applications. Because of its unusual properties asbestos fibers have since found their way into all aspects of our lives.

Much controversy surrounds the recognition that asbestos poses a serious health hazard. The first mentioning of potential health hazards associated with asbestos has been attributed to "Plinius" in the first century A.D. (Anderson 1902). In the late 1890's and early 1900's numerous reports appeared linking health problems to the asbestos industry. However, warning labels did not appear on asbestos products until 1964 (Weiss 1986). According to Antman

(1986) the recognition of pulmonary problems in workers from asbestos textile mills dates back to 1898. Because of the unsanitary conditions in most industries during that time period the asbestos health hazards could not be isolated until the early 1930's when the link between asbestos exposure and asbestosis was clearly established. Documents by Cooke (1924) and Merewether and Price (1930) have been attributed by Lee and Selikoff (1979) and Antman (1986) as keys to establishing scientific evidence of health risks associated with asbestos exposure. Since there is a very long latency period between exposure and the development of cancer the link between asbestos and lung cancer was not recognized until the late 1940's and early 1950's. Weiss (1986) claims that until the early 1960's there was insufficient scientific evidence to conclusively establish the health hazards associated with asbestos, and this is 20 years after the great increase in the use of asbestos.

1.4   Medical and biological problems

Asbestos fibers are known to be detrimental to humans, animals and plants. The response to exposure is complex leading to serious health problems in all three. Physical effects are thought to be the causes of ill health in man and some animals, while chemical effects are deemed critical to plant health.

1.4.1 Asbestos and human health.

There are a number of health issues associated with asbestos exposure. The most common medical problems are: pleural thickening and calcification, pulmonary fibrosis or asbestosis, lung cancer, mesothelioma and gastrointestinal cancer.

Pleural thickening and calcification is a nonmalignant formation of plaque in the lining of the chest cavity, which is suggestive of asbestos exposure but generally does not impair respiration. Pulmonary fibrosis or asbestosis was the first health issue associated with asbestos exposure. It refers to severe scarring of the lung tissues as a result of prolonged exposure to asbestos. Fibrosis is rarely fatal but develops progressively and affects the respiratory functions of the lung. It can lead to suffocation and sometimes cardiac failure (Charlebois 1978) and it is well known that people with occupational exposure to asbestos have a much higher incidence of asbestosis (Shugar 1979, Committee on Non-

Occupational Health 1984).

It is now established that extended exposure to high levels of asbestos leads to increasing rates of cancer. Lung cancer relates to the development of tumours in the lungs and is the leading cause of cancer in a number of asbestos industries. There is also a synergistic effect between asbestos exposure and smoking. People exposed to asbestos in many industrial settings have an 8 fold increase in lung cancer incidence than the general population. This incidence rate increases by 92 times for people with occupational exposure and smoking habits (Shugar 1979).

Mesothelioma is a rare cancer that develops in the membrane lining of the pleura. The incidence of mesothelioma has shown to be unusually high in persons exposed to asbestos and appears to affect people in both occupational and non-occupational settings. The medical profession has given much attention to this type of cancer because it is one of the few types of cancer that appears to be unique and intimately linked with asbestos. Wagner et al. (1960) produced the first firm evidence linking mesothelioma with asbestos. The extent of the disease apparently is extremely variable and dependent on the type of asbestos and the duration and intensity of exposure. Occupational exposure alone does not exclusively account for mesothelioma (McDonald and McDonald 1977 and 1980, Davies 1984, Wolf et al. 1987), but the link between non-occupational exposure to asbestos and mesothelioma has yet to be firmly established. Several studies have shown that mesotheliomas are frequent in some areas with asbestos rich bedrocks (Baris 1980, Rohl et al. 1982, Magee et al. 1986, Wagner 1986). Initially chrysotile was considered the major causative agent in mesothelioma but this view has been challenged by many researchers (McDonald 1980, Committee on Non-Occupational Health 1984, Rossiter 1987, Wagner 1986, Dunnigan 1988). Crocidolite and amosite fibers are thought to be more important than chrysotile fibers in the development of mesotheliomas (Wagner et al. 1973). Recent publications by Wagner and Pooley (1986), McConnochie et al. (1987), Churg (1988), Dunnigan (1988), and Craighead (1988) clearly suggest that fibers such as tremolite, which are present as contamination in many chrysotile deposits, might act as possible carcinogenic agents. Wagner et al. (1988) suggests that in terms of biological effects chrysotile is likely the least harmful of all the asbestos fibers and Wagner and Pooley (1986) even go as far as to suggest that any mineral fiber with diameters of less than 0.25 um and length greater than 5 um are now suspected agents for mesotheliomas, and that fiber analysis of the lungs appears to be the only means of determining lifetime exposure. More evidence to this end was provided by Wagner et al. (1985), and Baris et al. (1987), who showed that erionite fibers are also linked to mesotheliomas. In spite of these claims at the present time chrysotile cannot be excluded as a causative agent of both fibrosis and cancer (Pooley 1976, Wagner and Berry 1969, Wagner 1986, Churg 1988) since pure chrysotile without any amphibole fiber content has not been found.

Craighead (1985) suggests that genetic makeup or environmental conditions might be determinants in the development of this type of cancer since it is relatively rare, and not all people with a history of asbestos exposure develop the disease. The problem of synergistic effects caused by the presence of benzo(a)pyrene, N-nitrosodiethylamine, and cigarette smoking further complicate such analysis (Pott 1980, Lafuma et al. 1980, Mossman and Craighead 1981). However, the consensus amongst medical researchers is that the fiber geometry and size are the most likely agents involved with the development of asbestos related cancer (Timbrell et al. 1971, Stanton and Layard 1978, Pott 1980, Wagner 1986). In addition, dose and fiber durability are also critical components. Evidence is mounting that low level exposure to short fibers is not harmful (Dunnigan 1986) and models of linear response to exposure are now questioned. Instead, exposure threshold models are gaining in acceptance (e.g. Begin et al. 1987).

A number of authors have recently suggested that lung tissue analysis is the only way to determine fiber durability and exposure levels. However, this is an oversimplification since there are many problems associated with lung burden studies. McDonald (1988) mentioned that it is difficult to determine causes of cancer and the rate of exposure from the fibers present in the lung at time of death. Also, according to the same author the availability and amount of post mortem material for tissue examination is scarce and since the fibers are unevenly distributed in the lung it is difficult to arrive at quantitative results on fiber concentrations and fiber type. Finally, it is well known that there are a number of potential synergistic effects in asbestos related cancer and this suggests that closely controlled case studies are needed to obtain more reliable results.

A number of studies have reported elevated rates of gastrointestinal cancer in people with occupational exposure to asbestos but in most cases the rates were only slightly higher than in the general population and well within the magnitude of diagnostic and investigator errors associated with epidemiological research (Morgan et al. 1985). Considerable attention was given to the gastrointestinal cancer topic in determining the effects of asbestos ingested fibers via drinking water. In an early study Kanarek et al.(1980) reported a correlation between asbestos concentration in drinking water and gastrointestinal cancer. Many additional studies on the same topic but in different areas have failed to show such a relationship (e.g.Polissar et al. 1982, Marsh 1983, Condie 1983). Cancer risks from ingestion of asbestos fibers are now considered minimal (Levine 1985, Commins 1986, Velema 1987). Levine (1985) suggests that a final decision on the gastrointestinal cancer risk associated with asbestos ingestion must await further studies. The currently available information is not sufficient to determine the actual risk and it is unlikely that the assessment techniques currently used will provide us with a clear answer in the near future.

Finally, a number of nonmalignant pleural diseases have been linked to exposure to asbestos and other natural fibers (Becklake 1982, Hillerdal 1985). There have been several reports of pleural calcification and pleural plaque formation in people exposed to asbestos and erionite fibers in non-occupational settings. People living in anthopholite rich areas in Finland (Kiviluoto 1960), asbestos rich soils in Bulgaria (Burilkov and Michailova 1970), chrysotile and amphibole rich areas in Cyprus, Turkey and Greece (Yazicioglu 1976, Yazicioglu et al. 1980, Constantopoulos et al. 1985, Wagner 1986, Langer et al. 1987), and chrysotile and tremolite rich bedrock in Corsica (Boutin et al. 1986) have all been identified as being stricken by the disease. The use of tremolite rich rock material for whitewashing houses in Greece and Cypress (Constantopoulos et al. 1987a, McConnichie et al. 1987) had similar effects on people.

From this brief review of medical problems associated with asbestos it is clear that considerable progress has been made in determining the effects of asbestos on human health. The topic is very complex and much additional research is needed to gain a better understanding of the causes and prevention mechanisms of many of the asbestos induced diseases.

1.4.2 Asbestos and biological health

It is well known that asbestos and associated minerals have a detrimental effect on plant growth and animal life. The impact is known as the serpentine factor (Brooks 1987) and refers to the infertility and toxicity of serpentine bedrock and its associated ecology. In the ecological literature the term "serpentine" is used in many different ways and this makes a review very difficult. In a strictly mineralogical sense serpentine refers to the three minerals chrysotile, lizardite and antigorite, all of which have the same chemistry, but chrysotile is acicular while the other two are massive and non-fibrous. Unfortunately the term serpentine is used by many botanists to refer to the plants grown on all ultramafic rocks (Brooks 1987). This is very confusing since ultramafic rocks can be made up of more than olivine and pyroxene minerals which are the basic building blocks of serpentinitic rocks. A detailed discussion on this topic is provided in Chapter 2.

Worldwide, the ecology developed on serpentinitic bedrock is highly peculiar and often in stark contrast with the surrounding areas. The flora is poorly developed, the number of species limited, some plants are often chlorotic and stunted, and the areas affected by this parent material take on a desert like appearance.

The unusual ecology associated with serpentine rocks was noted in early history (Brooks 1987, Proctor and Woodell 1975 Krause 1958). Serpentine barrens are referred to as botanical islands and many plant ecologists have used these to study plant evolution. Stebbins (1942) and Brooks (1987) suggest that lack of competition or adaptation to the atypical environment has resulted in the establishment of these novel plant communities. Three major reasons are generally given for the poor performance of plants on such rock materials: metal toxicity, Ca/Mg imbalances, and deficiencies of basic plant nutrients. Proctor and Woodell (1975) suggested that the coarse nature of many serpentine soils might lead to moisture problems and this could be an additional factor complicating such investigations.

In contrast to medical research, I know of no study which investigates the possible penetration of mineral fibers into plant tissues and its effect on plant growth, distribution and vigour. Is plant root development restricted because of the abrasive nature of the asbestos fibers and do fibers pierce tissue during plant growth ? It is somewhat surprising that little research has been carried out on this topic considering the overwhelming evidence in the medical literature that fibers penetrate and pass through the human system and that the physical properties play a key role in the development of ill effects on human health. One of the reasons for the lack of research is that only a proportion of serpentinitic bedrock is in the form of fibrous chrysotile and its quantitative physical analysis is tedious and labour intensive. Very few papers in the field of plant ecology that deal with serpentine have reported the presence of fibrous minerals in the soil substratum, yet it is believed that chrysotile and to a lesser extent amphibole fibers are present in almost all serpentinitic materials.

Relatively little attention has been given to the effects of serpentine and asbestos on soil animals, micro-organisms and wildlife. The subject will be discussed in a later chapter but there is evidence to suggest that both the fibrous and non-fibrous forms of serpentine materials have adverse effects on animal ecology. Some of it is clearly related to scarce plant supply but the potential physical effects of fibers on the animals cannot be excluded.

1.5 Environmental Considerations

On a world scale asbestos rich rock formations are relatively small (< 1%). Commercial deposits are restricted to a handful of countries but ultramafic serpentinitic bedrock and other asbestos rich rock formations occur on almost all continents. Brooks (1987), in his excellent review of serpentinitic plants, has shown how plants can be used for geochemical prospecting and for identifying serpentinitic rocks. From the more than 1000 references cited in his book it is obvious that the natural distribution of serpentinite and its chrysotile components is widespread. The potential human health hazards associated with this material are dependent on the length and concentration of exposure and a host of genetic and environmental factors which have yet to be clarified by medical researchers. Natural exposure to asbestos is considerably more difficult to analyze than occupational exposure because the exposure levels are generally much lower. Many natural processes such as weathering, mixing of materials by glaciation, volcanic ash deposition, alteration by sedimentation and erosion affect the dominance and quality of asbestos in the natural environment. Given the complex mineralogy and the multiplicity of processes that can modify the asbestos rich residual conditions, it is difficult to assess the environmental health hazards. This is further complicated by the fact that asbestos fibers are redistributed within the environment by water and wind, and this further diffuses the assessment of ill effects on ecology and human health.

From the many studies of occupational exposure we know that asbestos is one of the most important natural carcinogens and a substance that has caused a lot of ill health. It is only in the last 10 - 15 years that researchers have focused on potential ill health effects from non-occupational exposure to asbestos. Progress has been made but many issues have not yet been resolved.

Exposure to asbestos fibers ingested via drinking water has, with the exception of one case (Kanarek et al. 1980), proven to be harmless since no ill effects could be documented in a large number of test cases in many areas of North America and Europe (DHHS Committee 1987). In contrast, people living in the vicinity of asbestos mines and in areas with asbestos rich bedrock have proven to be adversely affected in a number of cases (e.g. Burilkov and Michaelova 1970, Baris 1979, Bazas et al. 1981, Constantopoulos et al. 1985, 1987a&b, Boutin et al. 1986).

The adverse effects of asbestos on plant development and soil biota have been documented in cases of mine reclamation (Moore and Zimmerman 1977,1979, Meyer 1980, Perry et al. 1987) and in natural settings dominated by serpentinitic bedrock (Brooks 1987).

Effects of asbestos on animals and plants in the natural setting are currently of considerable interest. Belanger et al. (1986) has provided evidence of health problems in aquatic biota exposed to asbestos. Schreier and Timmenga (1986) have shown that earthworms are adversely affected by chrysotile fibers, and evidence exists that the activity of micro-organisms is significantly reduced in asbestos material (Proctor and Woodell 1975). A much greater research effort is needed in this area and it is hoped that this book will stimulate asbestos related environmental studies.

1.6 Topics to be addressed in this book

Given the complexity and diversity of asbestos one chapter will be devoted to a review of the properties, mineralogy, distribution and analysis of asbestos fibers. The importance of asbestos in the natural environment will be discussed in the following four chapters. Asbestos in the aquatic environment is the title of Chapter 3 which addresses the questions of sources, concentrations, transportation and deposition of asbestos fibers in the aquatic environment. It will also deal with fiber durability and effect on water quality. The impact of fibers on the aquatic biota, the use of asbestos rich water for irrigation and the effects of asbestos rich water consumption on human health are the main subjects to be reviewed.

Chapter 4 is devoted to asbestos in the soil environment. It discusses the genesis of soils from asbestos rich bedrock, the characteristics of such soils, and their adverse effects on plant growth and soil biology. How to deal with asbestos waste, how to modify asbestos rich soils to reduce hazards and to improve the growth environment are the subjects to be covered.

Chapter 5 is a review of the problems associated with plant growth on asbestos rich soils. A review of plant adaptation, modification and tolerance to asbestos will be provided. Plant species and plant communities indicative of such materials and non-endemic species and commercial crops useful for reclamation of asbestos rich or asbestos contaminated areas will be identified.

Chapter 6 addresses special environmental concerns involving recreational activities in areas naturally rich in asbestos, the use of asbestos rich materials for road fill and housing construction, dredging of asbestos rich sediments, and general problems relating to the disposal of asbestos in the environment.

Finally, a brief review of the major findings is provided in the epilogue and includes suggestions for further research on the topic of asbestos in the environment.

CHAPTER 2

PROPERTIES, MINERALOGY, DISTRIBUTION AND ANALYSIS

2.1 INTRODUCTION

Asbestos fibers are presently used in virtually hundreds of products and applications. This is largely a result of the novel properties inherent in the material. The physical, chemical and mineralogical properties have all contributed to making asbestos one of the most useful and, at the same time, one of the most hazardous natural substances in the world. To gain a better understanding of the impact of asbestos on the environment a brief review of its characteristic make-up is provided. This is followed by a discussion of source areas and distribution of asbestos fibers in the environment. The last section of this chapter presents a review of the analytical methods used to quantify asbestos. Publications by Spiel and Leineweber (1969), Coleman (1977), and Hodgson (1986) are excellent documents on properties and mineralogy. Riordon (1981), Ross (1984), Harben and Bates (1984), provide considerable information on asbestos occurrence and distribution, and Gravatt et al.(1978), Chatfield (1979 & 1986), and Chisholm (1983) provide basic information on asbestos fiber analysis.

2.2 PHYSICAL PROPERTIES

The most apparent feature of asbestos is its fibrous form and although many other mineral fibers exist the asbestos fibers are unique in form and characteristics. There is no single property that can be used to differentiate all types of asbestos. Its wide industrial use is primarily attributed to the novel physical properties of the asbestos minerals. And, on the other hand, some of the physical properties have also been identified by medical researchers as being the critical health hazard factors associated with the inhalation of the fibers. A summary of the major differences in physical properties between the different asbestos fibers is provided in Table 2.

TABLE 2

Overview of physical properties of asbestos fibers

Variables	Chrysotile	Amosite	Crocidolite	Anthophyllite
Fiber Length(um)	0.2 -200	0.4 - 40	0.2 - 17	
Fiber Diameter (um)	0.03-0.08	0.15-1.5	0.06-1.2	0.25-2.5
Surf. Area (m^2/g)	10 - 27	1-6	2-15	
Fiber Type	hollow tubes	-------- solid rods --------		
Tensil Strength(N/mm^2)	1000-2300	400-2100	2300-6900	apr.50
Magnetic Field	P	P	P & N	
Magnetic Susept. (emug Oc x 10^{-6})	5.3	78.7	60.9	14.3
Magnetic content(%)	0.5-2.0	0	3.0-5.0	0
Termal Stability(^{o}C)	600	600-800	600-800	
Max.Loss Ignition(^{o}C)	870	870-980	650	980
Surface Charge (mV)				
(neutral)	+93	-10	-10	
(alkaline)	-20 to -100	0	0	
Density (g/cm^3)	2.4-2.6	3.1-3.3	3.2-3.3	2.85-3.1
Filtration Properties	Slow	Fast	Fast	Medium
Moh's Hardness (1-10)	2.5-4.0	5.5-6.0	4	5.5-6.0

2.2.1 Fiber dimensions, shape and surface area

When viewed in the transmission electron microscope it is evident that asbestos fibers are arranged in bundles of tiny fibrils of various sizes, degree of peeling and openings. Bundles of fibrils may have length up to one cm and the individual fibrils share a common crystal growth direction. However, the structure of individual fibrils and their arrangement within the fiber differ amongst the various types of asbestos. As shown by Harington et al. (1975) there is a good relationship between fibril length and macro-bundle length (Figure 2). Usually, individual fibril length varies from 0.2-200 um. Larger fibers are most useful in industrial applications but are also more durable, hence more biologically hazardous. Most industrial grade chrysotile is in the size range of 1-2mm but there is a tendency for fibrils to break off or peel off by physical wear. Airborne dust in a number of industrial settings showed that the majority of asbestos fibers were smaller than 5 um in length (Gibbs and Hwang 1980). Chrysotile fibers were generally smaller than amphibole fibers and crocidolite fibers were usually smaller than amosite, anthophylite and tremolite fibers (Timbrell et al. 1970. and Langer et al. 1974). Amosite is unsurpassed in fiber length as compared to the other asbestos fibers, but it is somewhat more brittle. Usually amphibole fibrils are more rigid and parallel-sided than chrysotile fibers. In general, the length and width of fibers is very much dependent on the type of mineral fiber (Siegrist and Wylie 1980) and the geologic source area.

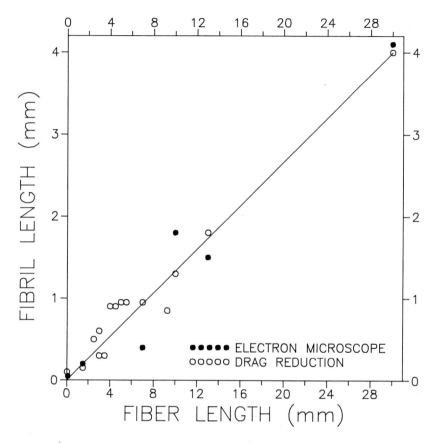

Fig 2. Relationships between asbestos fibers and fibrils (after Harington et al. (1975) Advances in Pharmacology and Chemotherapy, with permission).

Fibril diameters are usually small, reaching values of 0.03-0.06 um in chrysotile and 0.07 - 1.0 um in crocidolite, depending on sources and processing. The diameters of amphibole fibers are usually greater than their chrysotile counterparts and this results in fibers that are somewhat stiffer and less flexible. As shown by Whittaker and Wicks (1970), and Jefferson et al.(1978) the radius of curvature of chrysotile fibers is dependent on the degree of isomorphic substitution and the type of ion participating in the substitution. Nickel and cobalt substitutions result in decreased fiber diameters while aluminum substitutions result in flatter fibers (Chisholm 1983). As shown by Pooley (1986) diameters of crocidolite fibrils are generally smaller that corresponding samples for amosite but fiber width and fiber length are often positively correlated.

The length to diameter ratio (l/d) was found to be one of the useful identifying features of asbestos (Siegrist and Wylie 1980). All asbestos fibers have l/d ratios greater than 3:1, while most other natural mineral fibers have ratios smaller than 3:1. Most asbestos samples have skewed distribution and as shown by Siegrist and Wylie (1980) asbestos fibers have longer and more varied length, thinner and more varied widths, and greater and more varied aspect ratios than non-asbestos fibers. Following the theory that long and thin fibers are more carcinogenic than short and fat fibers, Stanton et al. (1977), Pott (1978) and Stanton et al. (1981) have given graphic information on length/diameter relationships and biological activities (Leineweber 1980). The positive relationships between fiber length and diameters have also been used by Virta and Segretti (1987) and Wylie et al. (1987) to develop models to predict fiber size distribution and fiber population by index particles in environmental and medical studies.

There are many critical problems associated with the measurement of fiber morphology. Some of these are origin, sample preparation, sample history (Gibbs and Hwang 1980), measuring method and instruments used (Chatfield 1979, Siegrist and Wylie 1980) and statistical uncertainties (Leineweber 1978, and Virta and Segreti 1987). This makes modelling of fiber geometry difficult.

As shown by Yada (1967) chrysotile fibers are hollow and have curly cylindrical structures due to the mismatch of SiO^4 and $Mg(OH)_2$ sheet structure. There is some doubt whether or not such fibers are hollow or filled. Amorphous materials exist in the fiber core and some of the contaminating minerals such as magnetite and brucite are often intergrown into the tubular or coiled structure. In contrast, amphibole fibers are solid rods with various ellipsoidal cross-sections.

The pore size distribution of the fiber matrix has been investigated in conjunction with specific surface area measurements (ssa). It is a measure of surface area per unit weight and expresses the fineness of the material. Nitrogen adsorption techniques are used to measure ssa, but the interpretation of the results is complicated because the degree of defibrillation and the degree of packing of fibrils into bundles greatly influence such measurements. Hodgson (1986) has shown that chrysotile generally has a large specific surface area, reaching values of up to 27 m^2/g. Crocidolite, with up to 15 m^2, and amosite with 6 m^2/g have considerably lower values.

2.2.2 Tensile Strength, Thermal and Magnetic properties

Tensile strength is of significant importance in industrial applications and Hodgson (1979) has shown that short fibers have higher tensile strength than long fibers. Chrysotile and crocidolite fibers have generally higher tensile strength than the stiffer amosite fibers.

The non-combustibility of asbestos minerals and their capacity to resist heat have contributed to the wide use of asbestos as insulation material. The mineral structure of chrysotile can be maintained up to 500-600° C with the loss of small amounts of water. Near 700° C there is a general structural breakdown where chrysotile is converted to fosterite. Structural alterations occur in crocidolite below 400° C and in amosite above 400° C but at 700° C amphibole type structures are still intact (Le Bouffant 1980). Above that temperature amphiboles are usually converted into pyroxene.

Magnetic properties of asbestos materials have been investigated because such properties are undesirable for a range of industrial applications related to electrical insulation. Chrysotile generally shows the lowest magnetic susceptibility of the asbestos minerals (Rendall 1980) while the more iron rich amphibole fibers, amosite and crocidolite, have the highest magnetic susceptibility. Badollet and Edgerton (1961), Berger (1965) and Schwarz and Winer (1971) have shown that most asbestos samples contain pure magnetite and Ni_3Fe but the separation of magnetite by magnetic forces has proven to be difficult particularly when dealing with the fine fiber fraction. De Waele et al. (1984) have shown that magnetite can be intergrown or can be present as a surface impurity and isomorphic substitution of iron for magnesium and silicon can also influence magnetism. Timbrell (1975) has characterized the magnetic properties of UICC asbestos samples and described a method of identifying asbestos fibers using parallel or normal to magnetic field alignments. By differentiating between paramagnetic and ferromagnetic components Stroink et al. (1985) were able to separate fibers contaminated with magnetite (ferromagnetic) from fibers subject to high iron isomorphic substitution for magnesium and silicon. Gale and Timbrell (1980) suggested that enough differences exist in magnetic alignments between different fibers that the magnetic techniques in conjunction with measuring the light scatter pattern could be used to improve current sampling methods. Willey (1987) suggests that such a method coupled with the measurement of magnetic field strength can be used to differentiate amongst respirable asbestos fibers and between asbestos and non-asbestos fibers for airborne sampling. However, Cressey and Whittaker (1984) showed that a separation of amphiboles is more difficult to make since crystallographic control around the principle magnetic axes could not satisfactorily be explained and the influence of magnetite on such alignment is questioned.

2.3 Chemical Properties

Because of the defects in mineral structure and isomorphic substitutions chemical analysis is usually not sufficient to differentiate between asbestos and non- asbestos minerals. Usually a combination of microscopic, mineralogical and chemical analysis is needed to quantify these minerals.

2.3.1 Major Elemental Composition

Complete elemental analyses of serpentine minerals have been provided by many authors, and site location, purity, mineral history and mineral structure are important factors determining differences in chemical constituents. Much more is known about the chemistry of chrysotile asbestos than about amphibole asbestos. There are some chemical differences between the fibrous and non-fibrous serpentine minerals. Whittaker and Wicks (1970) have shown that the structure, conditions of formation, and type and amount of substitution and contamination are responsible for chemical differences. Antigorite has generally higher SiO_2, lower MgO, and lower H_2O^+ values than chrysotile and lizardite and is usually formed under reducing conditions, while lizardite develops under oxidizing environments. This results in lizardite having more Fe^{3+} than antigorite, while the Fe^{2+} content is greater in antigorite. The iron content in chrysotile is generally lower than in the non-fibrous antigorite and lizardite minerals (Whittaker and Wicks 1970, Wicks 1979). Chrysotile and lizardite are considered polymorphs and differ slightly from antigorite in major elemental composition (Hodson 1986). Some analytical results of sepentinitic mineral and rock analyses are provided in Table 3.

TABLE 3
Major elemental constituents in serpentine minerals

	Chrysotile	Lizardite	Antigorite	Dunite & Peridotite Rock
	(1,2,4,5,6)	(1)	(1,3,4,5)	(7,8)
SiO_2	38 - 42	42 - 44	42 - 45	36 - 40
Al_2O_3	0.1 - 2.0	0.1 - 2.2	0.5 - 1.3	0.2 - 0.7
Fe_2O_3	0.1 - 5.0	0.5 - 1.3	0.1 - 1.7	2.8 - 6.0
FeO	0.1 - 3.0	O - 0.1	0.3 - 1.0	4.8 - 4.8
MgO	38 - 43	40 - 43	39 - 43	44 - 37
CaO	0 - 2.0	0 - 0.2	0 - 0.1	0.2 - 0.7
Na_2O	0 - 1.0	0 - 0.1	0 - 0.1	0.05 - 0.2
MnO	0 - 0.05	0 - 0.7	0 - 0.13	0.1 - 0.2
H_2O^+	12 - 14	12 - 14	12 - 14	8 - 11

Sources: 1) Deer et al. 1962
2) Rodrique 1984
3) Evans et al. 1976
4) Page and Coleman 1967
5) Oterdoom 1978
6) Hahn-Weinheimer and Hirner 1975
7) Trescases 1975
8) Krause 1958

Impurities and ion substitution are a serious problem in chemical analysis of all asbestos minerals. This is even more the case with amphibole minerals which have been referred to as wastebaskets in mineralogy. The variety of chemical composition is enormous and may include components of all the common mono-, di-, and trivalent metals. Chemical analysis is made particularly difficult because many other minerals are intimately intergrown with the fibers.

Some of the literature data on elemental composition of amphibole minerals is provided in Table 4.

2.3.2 Minor Elemental Composition
It is well known that serpentine minerals have unusual trace metal concentrations. Numerous authors have reported large

quantities of nickel, chromium and manganese in different asbestos minerals but as mentioned by Proctor and Woodell (1975) the variability in trace metals is large and comparisons are difficult to make because a wide range of analytical methods are used to quantify trace metal concentrations. Total elemental analysis is generally reported in medical and mining studies, while extractable, available and exchangeable trace metal values are often reported by soil and plant scientists. Numerous studies have investigated the influence of trace metals in biological activity. Although it is now the consensus of medical researchers that trace metals are unlikely to cause asbestos malignancy, the role of trace metals remains a controversial subject. Certainly the uptake of some of the trace metals from asbestos rich soils and bedrock by plants is of relevance since it appears to influence plant growth and vigour (see Chapters 3 & 4). In addition, some soil and aquatic animals living in such environments have shown elevated trace metal content and some of the trace metals can be used as a tracer to document influence and movement of asbestos fibers in the ecosystem.

TABLE 4

Major elemental constituents in amphibole minerals

	Crocidolite (1,2,4)	Amosite (1,2,4)	Anthophyllite (1,2,3,4)	Actinolite (1,2,3,4)	Tremolite (1,2,3,4)
SiO_2	44 - 56	48 - 53	53 - 60	49 - 56	55 - 60
Al_2O_3	0 - 1	0 - 1	0 - 5	0 - 3	0 - 3
Fe_2O_3	13 - 20	0 - 5	0 - 5	0 - 5	0 - 5
FeO	13 - 21	34 - 47	3 - 20	5 - 15	0 - 5
MgO	0 - 13	1 - 7	17 - 34	12 - 20	20 - 26
CaO	0 - 3	0 - 2	0 - 3	10 - 13	10 - 15
K_2O	0 - 1	0 - 1	0 - 1	0 - 1	0 - 1
Na_2O	4 - 9	0 - 1	0 - 1	0 - 2	0 - 2
H_2O^+	2 - 5	2 - 5	1 - 6	1 - 3	1 - 3

Sources: 1) Rodrique 1984
2) Spiel and Leineweber 1969
3) Whittaker 1979
4) Ney 1986

The trace metal content in high grade asbestos fibers (Union International Contre le Cancer - UICC - Standard reference samples and industrial grade mineral fibers) are provided in Table 5.

TABLE 5

Trace metal content in Serpentine and amphibole asbestos

	Co	Ni	Cr (ppm)	Mn	Fe (%)	Analytical Method & Ref.	
Chrysotile:							
UICC	45- 54	795-1445	316-1390	366- 443	0.88-1.14	HF/AA	(1)
UICC	40- 41	700- 880	460-1400	360- 460	2.2 -2.8	HF/AA	(2)
UICC	43- 54	802-1482	317-1378	231- 444	0.6 -1.24	HF/AA	(3)
UICC	45- 55	802-1482	317-1390	393- 480	0.6 -2.6	NAc	(4)
UICC	46- 55	990-1250	490-1390	393- 480	1.7 -2.6	NAc	(5)
Canadian	36- 78	299-1187	202- 771	325-1065		HF/AA	(6)
Canadian	39-110	550-2600	380-1200	1.8- 5.0		NAc	(5)
Canadian	44-110	330-1820	317-1200	420- 630	1.2 -4.8	NAc	(4)
Canadian	38- 60	917-1097	733-1369			NAc	(7)
Canadian	19- 43	63- 389	5- 79	145- 541	0.58-1.34	HCl/AA	(10)
African	54- 55	1360-1480	1378-1390	393- 450	0.6 -1.7	NAc	(4)
African	<100	1200-2900	600-1700	300- 800		NAc	(5)
Others		90-1700	40-1200	150- 740	0.95-1.68	NI/AA	(11)
Others	19- 49	493-1064	273- 919			NAc	(8)
Others		900	600	230	0.56	XFL	(9)
Crocidolite:							
UICC	2- 10	12- 58	16- 120	820-1320	15-33	NAc/AA	(2)
UICC	10- 12	0- 8	17- 20	833- 842	14-15	HF/AA	(3)
African	0- 7	<100	0- 20		20-26	NAc	(5)
African	<100	<100	<100	100- 300		NI	(7)
Amosite:							
UICC	7- 12	34- 58	32- 120	13600-15000	15-28	NAc/AA	(2)
UICC	11- 13	33- 35	31- 33	13347-13690	14-15	HF/AA	(3)
African	<100	<100	<100	1400-14800		NI	(7)
Others		300	2200	200	15.5	XFL	(9)
Anthophyllite:							
UICC	16-24	217-414	536-584	545- 986	13-20	HF/AA	(3)
Others		450	3000	2900	6.1	XFL	(9)
Tremolite:							
Others		700	1700	1600	3.5	XFL	(9)

1. Timbrell et al. 1968
2. Roy-Chowdhury et.al.1973
3. Cralley et al. 1968
4 Morgan et al. 1973
5. Holmes at al. 1971
6. Lockwood 1974
7. Cralley et al. 1967
8 Teherani 1985
9. Upreti et al. 1984
10. Barbeau et al. 1985
11. Reimschussel 1975

Techniques used:

HF/AA = Hydrofluoric acid digestion & Atomic Absorption Spectr.

NAc = Neutron Activation Techn.

XFL = X-Ray Fluorescence Techn.

HCl/AA = HCl digestion and Atomic Absorption Spectrometry

NI = Technique not identified

In comparison with most other non-metallic minerals in the environment the nickel, chromium and manganese values in many of these asbestos samples are exceptionally high. A number of studies have embarked on determining the sources and site locations of trace metals in the asbestos structure. As mentioned above magnetite is often present as an impurity in most asbestos specimens. Separating the magnetic from the non-magnetic fraction and subsequent trace metal analysis has shown that iron is only partially associated with the magnetic fraction (Reimschussel 1975). The non-magnetic rock and fiber fraction often contains iron values which exceed those found in the magnetic fraction, suggesting that a major part of the iron is isomorphously substituted for magnesium and silicon in the chrysotile mineral structure. Magnetite is often closely intergrown with the mineral fibers and cannot entirely be removed magnetically (Schwarz and Winer 1971, de Waele et al. 1984). This implies that the amount of isomorphically substituted iron is probably somewhat lower but nevertheless significant.

The majority of nickel was found in the magnetic fraction and is attributed to have replaced iron to form nickel-iron compounds such as awaruite. Some of the nickel in all chrysotile samples has entered the mineral structure by replacing magnesium.

According to Reimschussel (1975) chromium is mostly in the metallic state but might also be included in the chrysotile structure. Finally manganese is the most abundant trace metal in nature and is present in mineral form.

Similar studies examined the distribution of trace metals after hydrothermal treatment and mechanical cleaning of chrysotile. Hahn-Weinheimer and Hirner (1977) concluded that Fe^{+3}, nickel, cobalt and manganese can occupy octahedral sites in chrysotile.

The role of trace metals in asbestos malignancy is still a subject of uncertainty, although Wagner et al. (1973) and Langer et al. (1980) have suggested that trace metals are no longer an issue since asbestos samples with little trace metal content have also been shown to be cancer producing. Also, the trace metal issue is of considerable environmental interest in attempts to revegetate mine waste material.

2.4 Solubility and Leachability

Chrysotile and amphibole asbestos have very different solubilities in neutral, acid and alkaline media. The resistance of asbestos fibers to solution attack has been of interest in industrial applications but more recently attention has focused on solubility and leachability of fibers with respect to biological activity and asbestos disposal. In several epidemiological studies

it was shown that malignancy in test animals is significantly
reduced if chrysotile fibers have been leached by acids prior to
administration to test animals (Morgan et al. 1977, Monchaux et al.
1981, Jaurand et al. 1984, and Harvey et al. 1984). In addition
durability of fibers in lung tissue is linked to malignancy, and
solubility is linked to rate of fiber removal from tissue. In the
case of asbestos disposal attempts to treat the asbestos waste with
acids to increase disintegration has met with mixed success.

2.4.1 Chrysotile Solubility

The kinetic work by Choi and Smith (1972) and Papirer et al.
(1976) showed that chrysotile attack by water is limited to the
brucite layer with solubility products similar to those of
magnesium hydroxide. Defibrillation greatly influences the rate of
magnesium losses from the chrysotile structure and the type of
treatment influences solution pH and solubility. Steady state
conditions were reached at pH 9.5 - 10.5. However, Chowdhury (1975)
noted that true equilibrium is unlikely to be established since
metal ions are released at steady state due to the presence of
impurities in all asbestos samples making it virtually impossible
to study the actual leaching kinetics of natural chrysotile.

Over the pH range of 7-10, magnesium is released faster than
silica indicating initial surface release (Lin and Clemency 1981,
Bales and Morgan 1985a). The reaction is pH dependent and the
solubility in water is low. Using undisturbed and milled fibers
Harris and Grimshah (1975) showed that only 3% of all magnesium was
removed by water from the undisturbed fibers while 80% magnesium
was removed from the milled samples.

There are great differences in acid resistance and solubility
between chrysotile and amphiboles. If the acid conditions are in
excess and temperatures high chrysotile dissolves very rapidly,
but the rate is also dependent on type and strength of acid used,
wettable surface area and porosity of the fibers, and the amount
of interfibrillar contamination (Monkman 1971, Atkinson and
Rickards 1971, Morgan et al. 1971, Goni et al 1971, and Bleiman and
Mercier 1975). On the basis of these studies magnesium removal by
HCl is proportional to the square root until 2/3 is dissolved and
this suggests that about 1/3 of magnesium in chrysotile is in a
different less-reactive form (Barbeau 1979).

Oxalic acid appears to be the most reactive organic acid
(Thomassin et al. 1980) and diffusion of $Mg^{2}+$ through the fibrous
silica gel appears to be the rate limiting step (Thomassin et al.
1977, Thomassin et al. 1980, and Verlinden et al. 1984). The
fibrous structure can remain even after all magnesium was leached
out of the structure but the fiber fragments are of a gelatinous
or colloidal nature (Fripiat and Faille 1966, Spurny 1982, Seshan
1983, Bellmann et al. 1986). Morgan et al. (1971), Swenters et al.
(1985) and Parry (1985) provided calculated dissolution rates of
up to 2.13 x 10^{-7} moles hr^{-1} kg^{-1} and suggest that the surface area
of the dissolved mineral, mass of solution, rate of transport of
H^{+} into cells and $Mg^{2}+$ and SiO_2 out of the cell are the dependent
variables. Organic acids such as acetic, maleic, malic, and
tartaric acids remove some two thirds of the magnesium present in
the fiber structure, while malonic, tartaric and oxalic acids
extract 85 -95% of all magnesium over 20 hours at $100°$ C and pH 4
(Hodgson 1986). Leaching experiments have also been carried out
with carboxylic acids to simulate landfill leachate and its effect
on asbestos waste (Baldwin and Heaseman 1986) and this topic will
be discussed in more detail in Chapter 4.

What is clear from these experiments is that chrysotile asbestos
can readily be leached in acid rich environments but in spite of
removal of large quantities of magnesium the fiber structure
remains for a considerable time. In this leaching process the
charges on the fibers are usually reversed.

2.4.2 Amphibole Solubility

Amphibole fibers are resistant to both acid and alkaline
conditions, as shown by Hodgson (1979 and 1986). Anthopholite and
tremolite are particularly resistant to acids and according to the
same author the rate of loss appears to be dependent on the removal
of iron from the structure. As shown by Chowdhury (1975) solubility
of crocidolite and amosite in water is minimal. Very fine fibers
of the crocidolite and amosite variety can be leached in boiling
acids over considerable time and the leached residue is semi-
fibrous amorphous silica. It is difficult to remove sodium from the
crocidolite structure with mineral acids (Hodgson 1986) and
crocidolite and amosite are more resistant to acids and sea-water
than chrysotile. Under normal conditions oxalic acid had no effect
and tremolite (Mast and Dryer 1987) and crocidolite (Bellmann et

al. 1986) and the number of long (> 5um) fibers did not decrease in lungs of rats, suggesting that dissolution is limited. The widespread distribution of low grade amphiboles fibers and their general resistance to acid leaching makes them more durable, hence posing more long term environmental problems.

2.5 Physio-Chemical Conditions

The surface properties of asbestos have been of considerable interest to researchers not only because of their ability to alter asbestos performance for industrial purposes, but also because of the possibility that they play a role in determining biological activity. Surface charge and surface chemistry are the topics of most interest.

2.5.1 Surface Charge

The magnesium content in chrysotile and the surface charge as determined by zeta potential have been linked to potential causes of toxicity. The zeta-potential has been measured on chrysotile by a number of authors (Martinez and Zucker 1960, Ralston and Kitchener 1975, Chowdhury and Kitchener 1975, Leight and Wei 1977, Bales and Morgan 1985b, Isherwood and Jennings 1985) but as shown by Jacobasch et al. (1985) such factors as chemical constituents, fiber structure, porosity, specific surface area, interaction energy with water and swelling behaviour in water all influence the determination of zeta potential.

Chrysotile asbestos fibers freshly suspended in neutral to alkaline conditions have positive charges. Once the outer magnesium hydroxide layer dissolves in acid media the surface charges become negative due to the dominance of silica and the adsorption of natural organic constituents. According to Chowdhury and Kitchener (1975) different specimens show distinctly different zeta potential. Chrysotile fibers with excess magnesium have strong positive charges, weathered samples are weakly negative, and acid leached chrysotile fibers have charges that are strongly negative. Suquet et. al. (1987) have shown that the weak surface charge of chrysotile is very localised and about 1/3 of the charge comes from magnesium vacancies near the surface of the fiber. Seshan (1983) suggests that due to leaching, chrysotile fibers become more silica like and the x-ray signal of crystallinity disappears with excessive leaching. According to Bales and Morgan (1985b) the adsorption of Mg^{2+} to the remaining SiO^- and $SiOH$ is insignificant but the adsorption of organic constituents increases over time.

In amphiboles no detectable changes were observed by acid leaching and zero point charge is generally negative for all amphiboles (Ralston and Kitchener 1975, Leight and Wei 1977). The magnitude of the zeta potential of crocidolite, amosite and anthopholite is about the same as unleached chrysotile but amphiboles have reversed polarity. Surfactants can be adsorbed to the fiber surfaces and this reduces the zeta potential. As noted by Leight and Wei (1977) such reductions are considerably higher in the amphiboles, crocidolite and amosite than in chrysotile. Many molecules can be adsorbed at the fiber surfaces. Constituents can be adsorbed chemically or by chemofixation, which is the incorporation of molecules into the micro-pore structure of leached fibers (Seshan 1983).

2.5.2 Surface interactions

Interactions of fibers with the physical, chemical and biological environment can change the fiber chemistry and crystalline structure of the fibers. Some fibers remain unchanged, some become partially leached and in some cases the majority of metals are removed from the fibers depending on the environment (Spurny 1981). The dynamic nature of these processes in different environments requires careful fiber analysis before and after each experiment using such techniques as scanning electron microscopy, electron microprobe analysis, mass spectroscopic analysis, and scanning auger spectroscopy (Spurny 1983, Pathak and Sebastien 1985, Verlinden et al. 1985, Jolicoeur and Poisson 1987).

The adsorption of selective molecules is pertinent because many have the capacity to alter fiber properties for industrial purposes and some might be possible causes for malignancy. With a better understanding of these processes it might be possible to artificially alter surface conditions to reduce the toxicity of asbestos fibers. Without altering fiber size, elemental content and fiber morphology, simple heat treatment resulted in lower toxicity in chrysotile and is according to Fisher et al. (1987) attributed to electron transfer. Adsorption studies have shown that toxicity is also reduced by adsorption of titanium (Cozak et al. 1983) and phosphorus (Khorami and Nadeau 1986, Khorami et al 1987).

In the latter case a coating of insoluble polyphosphate is formed at the fiber surface with oxychlorite and subsequent heat treatment.

Adsorption of organic surface agents causes dispersion or defibrillation of fibers and improves properties relevant to textile production, reinforcing plastics and rubbers, and improves the industrial use of oil fluids in drilling operations (de Waele et al. 1985). In general organic constituents are weakly adsorbed to chrysotile, and leaching of the Mg(OH) layer is needed to develop good organic-mineral bonding. Many leaching reactions are harsh and alter the fiber structure. Controlled removal of the brucite layer is possible with ethylene diamine tetracetic acid (EDTA) and this allows the modification of a relatively undisturbed silica fiber surface. Polycyclic aromatic hydrocarbon (benzo[a]pyrene) in asbestos has been of some interest because of its possible role as a carcinogen (Mossman 1983, Harvey et al. 1984). As shown by Gibbs (1971) organic constituents in chrysotile minerals are relatively small and can be the result of gas adsorption or contamination of packaging materials (Commins and Gibbs 1969).

Schiller et al.(1980) and Bonneau et al. (1986) have shown that both amphibole and chrysotile asbestos may have dual surface charge characteristics. The extremities of the fibers where the octahedral sheet emerges appear to have a weak positive charge while the lateral surfaces are negatively charged. This would increase the complexity of adsorption analysis.

2.6 Mineralogy and Geology

Amphiboles and serpentine minerals can occur in both massive and acicular form and it is usually the latter which are referred to as asbestos fibers. While there are many other mineral fibers, true asbestos fibers break up into smaller and smaller fibrils. A fibril is a single or twinned crystal of very small width and high aspect ratio, and their common crystal growth direction is only along the long axis of the fiber (Steele and Wylie 1981).

Amphiboles are made up of double chain silica tetrahedra structures, while serpentines are arranged in a sheet silicate structure. Amphibole and chrysotile asbestos have fairly complex mineral structures because of the many disorders and substitutions that occur within the mineral structure. Most asbestos minerals are confined to metamorphic rocks or ultramafic and sedimentary rocks with various degrees of alterations.

2.6.1 Amphibole minerals

Amphibole minerals are fairly abundant in igneous and metamorphic rocks and Whittaker (1979) has estimated that up to 20 % of the shield areas of the world contain some form of amphibole minerals. In most cases the amphibole minerals are massive and their common feature is cleavage along two sets of planes parallel to the crystallographic z-axis, with a 55° angle between the planes. Because of this cleavage amphibole minerals break into elongated particles that are needle like but are generally not considered fibrous unless they can be split into smaller and smaller fibrils.

As shown by Whittaker (1979) the amphibole structure provides many opportunities for interchangeability of elements. Amphiboles are also known for having many structural irregularities and as a result are particularly difficult to classify mineralogically (Veblen et al 1977). The interpretation of chemical formula by chemical analysis is equally challenging. There appear to exist structurally ordered and disordered phases in the single and double silica chain structures. The presence of chain width errors is, according to Veblen et al. (1977), the common denominator between all amphibole asbestos and he refers to these mineral structures as pyriboles. There are two types of faults that appear to contribute to the fibrosity of amphibole asbestos: stacking order faults and Wadsley defects. The stacking order faults are found in the repetition of the sequence of tetrahedral - octahedral layers in a two chain model. These basic structures form blocks and the sequence of blocks was shown to be variable in all fibrous amphiboles (Chisholm 1973, Veblen et al. 1977, Hodgson 1986). Wadsley defects refer to problems in the sequence of shear planes. Unequal sequences of silica sheets that occur between crystallographic shear planes result in gains and losses of cations and OH groups.

Suggestions have been made that the presence of aluminum in tetrahedral layers inhibits fiber formation but Whittaker (1979) also indicates that the conditions of crystal growth are very important in determining whether amphiboles develop in acicular or non-acicular form. Research by Jenkins (1987), Hutchison et al.(1975), Oterdoom (1978) and Skippen and McKinstry (1985) has provided some information on growth characteristics of several synthetic amphiboles, and Subbanna et al. (1986) elaborated on the fact that iron from the host minerals is accumulated in brucite minerals that are intergrown in anthophylite asbestos.

As indicated by Hodgson (1986) some progress has been made to develop a better understanding of amphiboles but the chemical and mineralogical analysis of amphiboles remains challenging.

The best crocidolite and amosite minerals are found in old Pre-Cambrian, iron rich, banded sedimentary rocks which are rich in silica, magnesium, and sodium and have undergone metamorphic transformations (Zussman 1979). The fibers are thought to be formed in a secondary process of recrystallization involving hydrothermal alterations. Anthophylites develop during metamorphic processes at high temperatures in association with ultrabasic rocks and gneiss, while tremolite is often associated with marble and some serpentine minerals (Oterdoom 1978).

2.6.2 Serpentine Minerals

Serpentine minerals occur in both massive and fibrous form and usually have layered sheet structures made up of silica tetrahedra and brucite type ($Mg(OH)_2$ octahedral unit cells. The chemical complexity is generally lower in serpentine than in amphibole minerals, but the sepentine structure is often more complex than that of amphiboles (Wicks 1979).

Single flat layered structures usually represent lizardite, corrugated layers antigorite, and cylindrical or spiral layers reflect chrysotile minerals (Wicks and Whittaker 1975).
Due to the impurities present in all asbestos, systematic chemical differences are difficult to detect between the three types of serpentine minerals (Deer et al. 1963). Lizardite is usually more variable because it has a greater capacity to accept substitutions. Variations in aluminum substitution in chrysotile appears to reduce the mismatch between the $Mg(OH)^2$ and the Si_2O_5 layer thus resulting in a flatter non-fibrous structure. According to Hodgson (1986) the structural defects and the degree of aluminum substitution in both serpentine and amphibole minerals might provide the key to understanding the presence of fibrous components in serpentine and amphibole minerals. Also some polymorphs are favoured by particular environmental conditions and according to Wicks and Whittaker

(1977) temperature, presence and absence of shearing, and extent of nucleation of antigorite are the key components determining their occurrence.

Chrysotile fibers are hollow and, as shown by Yada (1971), the voids between fibers are often filled by contaminants. Brucite and magnetite are the most common interfiber contaminants and in the serpentinization process they are usually enriched and contain more iron than either lizardite, antigorite, or parent olivine (Moody 1976, Whittaker and Middleton 1979).

Serpentine minerals are usually associated with ultramafic rocks which have been altered at different temperatures and in the presence of water. This alteration process, which is known as serpentinization, is widespread and as a result chrysotile asbestos is likely to be present in most serpentinized ultramafics in the world, but the quality and quantity are rarely sufficient for commercial exploitation. The most common mineral constituents of ultramafic rocks are olivine, pyroxene, and hornblend and the dominant rocks are dunite (mostly olivine), pyroxenite (mostly pyroxene), peridotites (olivine rich with some pyroxene), and harzburgite (mostly olivine and pyroxene with some plagioclase). The relative distribution of these basic minerals in most common ultramafic rocks has been discussed by Wyllie (1967) and Coleman (1977) and is illustrated in Figure 3.

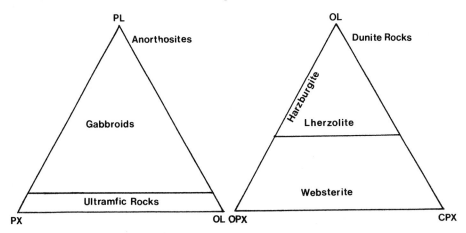

Fig. 3. Model proportions of minerals to differentiate ultramafic and serpentine rocks (after Coleman 1977, Springer Verlag, N.Y., with permission.

The serpentinization process is thought to be a two stage process in which the base rocks are altered by metamorphic processes and the asbestos fibers are formed in the second step involving water solutions which are likely of oceanic origin. Evidence of this was provided by Moody (1976) who indicated that boron is usually enriched in serpentine as opposed to the parent rock. The different reactions between dunite, pyroxenite and fluids, the temperature and pressure regime for hydration, the mineral assemblage and fluid composition all play a role in the serpentinization process (Moody 1976, Yada and Iishi 1974, Evans et al. 1976). Besides altered ultramafics chrysotile has also been found in serpentinized dolomitic marble.

Some chrysotile fibers are soft and flexible, others are harsh, brittle and inflexible. As noted by Wicks (1979) the relationship between these properties and mineralogy is still uncertain.

2.7 Geologic Distribution of asbestos

Given the importance of asbestos in industrial applications considerable efforts have been made to document the location and reserves of commercial grade asbestos. The geological formations which bear high grade asbestos are well known but relatively scarce. In contrast, non-commercial asbestos occurrences are frequent but poorly documented and this makes it difficult to assess asbestos in the environment. In addition asbestos fibers are very small and mobile and are readily redistributed by wind, water, ice, and gravity.

2.7.1 Distribution of Commercial Grade Asbestos

Traditional asbestos mining areas are the Eastern Township of Quebec, Canada, the Northern Cape Province and Transvaal in South Africa, and the Southern Ural Mountains in the USSR. Expansion in asbestos mining took place in the 1970's in response to the great demand as new industrial applications were found. However, economic recessions in the early 1980's and environmental health concerns have reduced the demand for asbestos ore considerably and a number of mines have recently closed and others are likely to close in the very near future.

As noted by Ney (1986) approximately 95% of all mined asbestos is chrysotile, 3% is crocidolite, 1.5% is amosite, and the remaining 0.5% are other asbestos varieties. In the early 1980's world production was estimated to be in the order of 5 million metric tons (Shugar 1979, Ney 1986) but production has since declined. Between 1947 and 1974 Canada supplied between 40-60% of the world asbestos, but this has now declined to less than 30% and the USSR is now the leading supplier (Shugar 1979). South Africa is supplying between 7-10% of asbestos and most of the industrial grade amosite and crocidolite asbestos originated from there. A list of producer countries and their potential capacity are provided in Table 6 below.

TABLE 6

Estimated production capacity for commercial grade asbestos

Country	Estimated Production Capacity (Tons)
USSR	3,100,000
Canada	1,500,000
South Africa	400,000
People's Republic of China	300,000
Zimbabwe	300,000
Brazil	200,000
Italy	200,000
USA	120,000
Greece	100,000
Australia	100,000
West Germany	90,000
Swaziland, Cyprus, India, Japan, Yugoslavia, Columbia, Turkey, etc.(each)	>50,000

Based on : Shugar (1979), Ney (1986),

Ross (1982) has provided a description of the major commercial deposits and their associated rock formation. He differentiates between: Alpine type ultramafics, stratiform ultramafic, and serpentinized dolomitic limestone as host rocks for major commercial grade chrysotile, and metamorphosed banded ironstone, and alpine type ultramafic rocks for commercial amphibole asbestos. A list of major commercial deposits is provided in Table 7 which summarizes essential geological information and literature sources describing the deposits.

2.7.2 Occurrences of Non-Commercial Asbestos

Low grade asbestos fibers occur in many locations around the world. However, to arrive at a worldwide distribution is difficult because the proper detection and analysis of asbestos is expensive,

Table 7

Some of the important commercial deposits of asbestos fibers

Country	Location	Asbestos Type	Rock Formation	References
USSR	S-Central Ural Sverdlovsk, Tuva & Kustanay Region	C	US	Harben & Bates (1984) Petrov & Znamensky (1981)
Canada	Eastern Quebec	C	US	Lamarche & Riordon (1981)
	N-E Quebec	C	US	Hanley (1987), Stewart (1981)
	Newfoundland	C	SP	Williams et al. (1977)
	British Columbia	C&T	SP	Burgoyne (1986)
USA	N-Central Vermont	C	P&D	Chidester et al. (1978)
	California	C	SP	Mumpton & Thompson (1975)
	Arizona	C&T	AL	Harben & Bates (1984)
	Georgia-Maryland	An&C	U	Ross (1982), Puffer et al.(1980)
	New Jersey	C&T	AM	Germine & Puffer (1981)
Yugoslavia	Croatia	C	SP	Harben & Bates (1984)
Greece	Macedonia	C	H&I	Harben & Bates (1984)
Cyprus	Troodos Mts.	C	US	Harben & Bates (1984)
South Africa	Transvaal	C	AS	Dryer & Robinson (1981)
	Transvaal Lyndenburg	A&Cr	BI	Ross (1982)
	N-Cape Province	Cr	BI	Dryer & Robinson (1981)
Swaziland	Northern Region	C	S&C	Harben & Bates (1984)
Zimbabwe	Eastern Bulawayo	C	D&P	Harben & Bates (1984)
Australia	New South Wales	C	H&D	Butt (1981)
Finland	Karelian Mts.	An	US	Ross (1982)
Italy	Western Alps	C&T	S	Ross (1982)
China	Various Locations	C&T	U&Do	Hodgson (1986)
Brazil	Goias State	C	D&P	Beurlan & Cassedanne (1981)

Asbestos Type:
C = Chrysotile
T = Tremolite
An = Anthophyllite
A = Amosite
Cr = Crocidolite

U = Ultramafic Rock
Se = SerpentineRock
P = Peridotite
AS = Altered Sedimentaries
H&I = Harzburgite & Iherzolite
Do = Dolomite
AL = Altered Limestone
BI = Banded Ironstone
SP = Serpentinized Peridotite
D = Dunite
AM = Altered Marble
S&C = Altered Schist & Carbonates
US = Ultramafic Serpentinite

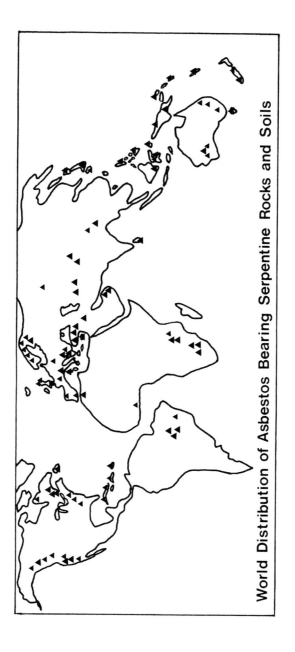

Fig. 4. Distribution of asbestos bearing serpentinitic bedrock in the world
(after Brooks 1987, Kruckeberg 1984 and Hodgson 1986).

time consuming and requires specialized equipment. Chrysotile asbestos is commonly associated with all serpentinitic rocks in the world and in spite of the massive literature on serpentinites (Brooks 1987) very few researchers have actually measured the amount of asbestos fibers in serpentinitic bedrocks. Given the enormous variability in serpentinites (Proctor and Woodell 1975) and other altered asbestos bearing rocks the task of determining exact asbestos fiber concentrations in rocks is virtually impossible. The assumption was thus made that all serpentinitic bedrock contains asbestos fibers, and data provided in Figure 4 gives a generalized representation of rocks which have a very high probability of containing some form of asbestos fibers. Details of the distribution of serpentinites on a sub-continent/country basis are provided by Brooks (1987). Kuryval et al. (1972) provided a map of asbestos distribution on a county basis in the USA, and Kruckeberg (1979 and 1984) provided information for California and the west coast of North America.

Only a portion of the asbestos bearing strata is directly exposed at the earth's surface, but man's activities and many natural processes are responsible for bringing asbestos to the surface and redistributing it throughout the environment.

Mining operations play a significant role in exposing asbestos fibers. In such operations it is not unusual to have an overburden to mineral ratio greater than 3:1 and this requires large amounts of overburden materials to be moved and deposited. Much of this waste material contains asbestos either by association or mixing during the mining process. Asbestos ore which is actually mined has to go through a screening and sorting process and another 50 - 80% of the ore is eliminated and returned to the site as waste material.

Much asbestos bearing rock has been quarried and is in frequent use for road fill and other construction purposes. Much asbestos used in cement and industrial application eventually returns as waste to land fills, and surface disturbance by different land uses such as agriculture, off road recreation traffic, construction activities and natural erosion all contribute to increasing the exposure and redistribution of asbestos at the land surface.

Because of their small size the fibers are very mobile and water, wind and ice can readily redistribute asbestos within the environment. Asbestos fibers are therefore present not only in

asbestos bearing bedrock but also in surficial materials
transported by glacial activities in the past and by wind, water,
and gravity. There is of course a dilution effect once the fibers
are transported and mixed and the overall concentrations may be
drastically reduced. In addition, weathering processes alter and
break down some of the fibers. The resistance to weathering and the
conditions and rates of breakdown vary greatly and will be
discussed in greater detail in Chapters 3 and 6.

2.8 Identification and Quantification of Asbestos

Asbestos minerals are defined as hydrated silicate fibers which
have high tensile strength, high flexibility, length to width
ratios of > 3:1, and are arranged in bundles of fibrils that can
easily be separated (Zoltai 1978). Several techniques are therefore
required to adequately quantify asbestos fibers.

In most western countries there is a legislation which limits
asbestos fiber exposure in the workplace. To uphold the legislation
monitoring programs and analytical methods had to be developed to
identify and quantify asbestos fiber concentrations in the
environment. While there is agreement on the limits of each
analytical technique the universal use of standard methods leaves
a lot to be desired. This is largely due to the high cost of
analysis, the lack of easy access to certain types of analytical
tools, and the uncertainty of whether to measure fiber morphology,
size, weight, elemental and surface composition, surface charges
and mineral configuration. A major analytical problem is that the
crystallinity and composition should not be altered during the
process of measuring the morphology, mineralogy and composition of
the fibers.

2.8.1 Morphological analysis

The definition of fibers, that they must have aspect ratios of
> 3:1 and must be longer than 5 um, is clearly insufficient because
more then 150 different types of minerals can form fiber-like
particles with the above mentioned ratio (Hodgson 1986).

Morphological analysis of fibers has traditionally been carried
out using contrast optical microscopy. This method, however, is
limited to particles with diameters greater than 0.4 um and is
unsatisfactory because it only provides information on the number
of fibers present without reference to the fiber origin or

composition. In addition, fibers with diameters smaller than 0.4 um cannot be detected, but are likely to be the most hazardous to human health. The reliability of this technique for asbestos fiber analysis has been questioned by Chatfield (1979) since it is used at the lower end of its detection limit and is thus subject to operator error. The method can be improved in combination with dispersion staining techniques where dyes are added to the samples and variances in refractive index are measured in different sample orientations (McCrone 1978, Ganotes and Tan 1980). The above mentioned resolution problem and the requirements of having to know the optical and morphological properties of all transparent substances in the same sample are drawbacks in the use of this technique for mixed environmental samples.

Transmission electron microscope (TEM) analysis is considered the most reliable method of fiber counting and morphological analysis (Chatfield 1979). However, to use this technology is tedious, time consuming and therefore expensive.

In both industrial and environmental settings fiber counting has been the most popular way of measuring asbestos fiber concentrations. Automated, laser based particle counter techniques are used but they are based on measuring the intensity of light scattered by the particles, hence they can size particles but cannot provide information on particle shape. Filter devices alone and in combination with image analysis are also used as techniques to determine fiber numbers in samples. These have been perfected by aligning the fibers according to their magnetic properties (Timbrell 1975) and sizing the fibers with computer based video imaging techniques.

By far the most reliable measuring technique is still the manual counting of fibers on transmission electron microscopes (TEM). This is done after samples are filtered through carbon coated Nuclepore filter, mounted on a copper grid and the filter paper dissolved with chloroform (Chatfield 1979).

2.8.2 Fiber Identification Techniques

IR Spectroscopy has been used successfully to differentiate between the mineralogy of the different fibrous particles. While the technique is successful in industrial applications it has proven to be of limited use in the analysis of airborne and other environmental samples since a minimum sample of 20 ug is needed for accurate detection of asbestos fibers (Gadsden et al. 1970).

Selective area electron diffraction (SAED) and energy dispersive X-ray analysis (EDXA) have been used in conjunction with TEM and SEM analysis and these two techniques have proven to be the most successful techniques for fiber identification. An example of a typical energy dispersive x-ray analysis of chrysotile asbestos is provided in Figure 5.

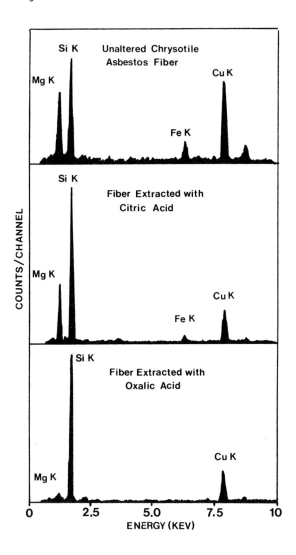

Fig. 5. Energy dispersive X-ray analysis (EDX) of unaltered, and acid leached chrysotile asbestos fibers.

Different asbestos fibers are known to have unique selective area electron diffraction patterns and thus give information on the crystalline structure. The hollow tubular structure of chrysotile give a particularly distinctive pattern if the fibers are neither thin nor thick and relatively free of contamination (Beaman and Walker 1978). The measurement of numerous fibers is required and even then it is often extremely difficult to differentiate between some of the amphibole fibers (Hutchison et al. 1975). Examples of SAED patterns for all different types of asbestos fibers can be found in the asbestos fiber atlas published by Mueller et al. (1975).

Energy dispersive X-ray analysis (EDXA) is useful because it provides information on the elemental composition of the fibrous material. Its sensitivity is limited to slightly less than 1% elemental composition and cannot be used to consistently differentiate amongst amphibole fibers.

Every one of these techniques has distinct limitations but used in combination the TEM/SAED/EDXA analysis provides the best means to quantitatively measure fiber concentration, identify asbestos from non asbestos fibers, and to determine compositional differences between asbestos materials (Asher and McGrath 1976, Beaman and File 1976, Stewart 1978, McCrone 1978, Chatfield 1979).

2.8.3 Sample Preparation

Sample preparation is most critical when analyzing environmental asbestos samples with scanning and transmission electron microscopes (STEM). The most common procedures involve filtration of samples with 0.1- 0.2 um size Nuclepore filters. Samples are prepared by concentration washing, carbon coating and filter extraction (Jaffe-wick method) with chloroform. Sample contamination and fiber losses during filtration are the main problems associated with sample preparation. Airborne environmental samples and tissue samples are best subjected to low temperature plasma ashing to remove heavy filter and tissue components. The remaining ash is then suspended in water, subjected to ultrasonic treatment to disperse the sample and then placed on Nuclepore filters and carbon coating as mentioned previously (Beaman and Walker 1978, Chatfield 1979).

Although there is no generally approved standard method the methods described by the National Institute of Occupational Health and Safety and the Environmental Protection Agency (EPA) are most widely used (Walton et al. 1976, Thompson 1976, Anderson and Long 1980, Chatfield et al. 1983). This involves the sample collection, deposition on filters, ashing and refiltration, clearing of filter, scanning and counting of samples (Lineweber 1978). Problems resulting from poor mixing, clustering of fibers, fragmentation of fibers, contamination of samples, and identification difficulties all contribute to the uncertainty in determining fiber concentration and fiber type. Leineweber (1978) suggested that description of sampling conditions, volume used for filtration, method of sample preparation, number of samples and grid cells counted, identification of fibers and fiber dimensions are all essential for determining the significance of the results obtained.

Many comparisons have been made between different laboratories using the same methods (Beckett and Attfield 1974, Brown et al. 1976, Gibbs et al. 1977, Fernandez and Martin 1986) and in spite of considerable efforts the main questions of accuracy, contamination and reproducibility remain. In order to improve the analytical accuracy between laboratories guidelines are still forthcoming (Cooney 1987) and, given the complexity of asbestos fiber analysis, the topic will remain highly controversial for many more years to come (Krantz 1987).

2.8.4 Standard Reference Samples of Asbestos

The working group on Asbestos and Cancer of the UICC (Union International Contre le Cancer) recommended in 1964 to prepare standard reference samples of the major commercial deposits of asbestos fibers. These included South African amosite and crocidolite, Finnish anthophylite, and Canadian and Rhodesian chrysotile. These asbestos sources made up the majority of commercially available asbestos and it was felt that detailed characterization of the fibers was needed given the variability inherent in asbestos samples in general. It was felt that such reference materials could do much to improve methodology and reproducibility of medical and biological tests involving asbestos. These reference samples have been described by Timbrell et al. (1968) and their chemical and physical characteristics have been elucidated by Timbrell (1969). These samples have served as reference standards in most analytical work and animal exposure experiments. Some 20 years have passed since the initiation of the standard reference samples and progress has been made in gaining a better understanding of fiber characterization. Unfortunately, the actual cause of cancer from asbestos fiber exposure has so far eluded us. What has been accomplished is more uniformity and reproducibility in test results with the use of these standard samples. But the specific focus is on industrial use and exposure and, given the great variation in asbestos, these standards are not entirely appropriate for examining environmental exposure to fibers. Nevertheless, they are serving as an essential data base for comparison purposes.

In recent years a new controversy has occurred since small amounts of tremolite fibers have been found contaminating all chrysotile samples. This is now putting in question whether and, if so, to what extent chrysotile alone is responsible for developing mesothelioma (Wagner 1986, Dunnigan 1988, Craighead 1988).

2.9 Summary

Asbestos fibers have some of the most novel physical and chemical properties of any natural substance. The physical properties of heat resistance, high tensile strength, and non-combustibility, and the selective chemical resistance to acid and sea-water have made these minerals exceptionally useful to man. In contrast, the physical properties also pose a most serious threat to human health, and the combination of physical and chemical composition make asbestos rich soils an undesirable environment for plant growth.

The physical, chemical, and physio-chemical properties of asbestos are of major significance in the environment. Long, thin and durable fibers appear to be most dangerous to human health and their physical quantification remains a tedious and expensive process requiring electronmicroscopy coupled with energy dispersive X-ray analysis techniques. The determination of the fiber mineralogy and chemistry is also complex, because of the many defects in the structure of the fibers caused by isomorphic substitutions. In addition most asbestos fibers are contaminated with other minerals and trace metals are particularly abundant. Other complicating factors are the physio-chemical properties and their interactions with organic matter.

The asbestos minerals are divided into chrysotile and amphibole asbestos fibers. The former is the most widely used and distributed fiber but can readily be leached in acid environments. In contrast the amphibole fibers are acid resistant and their mineralogical analysis is very challenging.

CHAPTER 3

ASBESTOS IN THE AQUATIC ENVIRONMENT

3.1 Introduction

Until the early 1970's little was known about asbestos fibers in the aquatic environment. It was generally believed that high fiber concentrations were restricted to industrial and mining activities involving asbestos. Efforts were made to determine fiber concentrations in drinking water supplies in both North America and Europe and from these studies it became evident that asbestos fibers were present in many drainage systems around the world. The aims of this chapter are to document natural concentrations of asbestos fibers in the aquatic system, compare them with industrial effluent sources, discuss the movement of fibers within the water systems, evaluate the effects of fibers on water quality, review what is known about the impact of water borne asbestos on aquatic and human health and discuss the potential impact of using asbestos rich water for different land uses.

3.2 Asbestos Fibers in the Water Supply

Because of their small size and low density, asbestos fibers can readily be transported by wind and water and thus enter a variety of environmental pathways. The pathways by which asbestos fibers are most frequently introduced into the aquatic system are shown in Figure 6 and it appears that non-point sources contribute far more asbestos to the overall water system than point sources. Precipitation acts as a collector and thus introduces fibers into the hydrological cycle by rain and snow. Stream and groundwater contact with asbestos-bearing bedrock and surficial materials seems to be the major source of asbestos fibers in the water supply and many natural streams have been shown to contain concentrations that are as high as or higher than effluent water from the asbestos mining industry.

Contribution from asbestos cement pipes and other construction materials used to transport and store drinking water is another substantial source of asbestos in the aquatic system. Vehicular traffic and road run-off are also considered a major source of asbestos fibers. It is obvious that both point and non-point

sources contribute to the asbestos fiber concentrations in stream
water, lakes and water supplies.

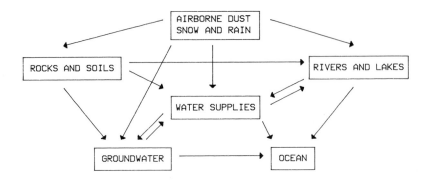

Fig. 6. Asbestos fiber pathways in the aquatic system

3.2.1 Asbestos Fibers in Drinking Water

Some of the early indications of asbestos fibers in drinking
water were reported by Cunningham and Pontefract (1971) in Quebec
and Ontario and by Cook et al. (1974 and 1976) in the Lake Superior
area. Asbestos fiber concentrations of $10^6 - 10^8$ fibers/litre
were reported and this led to numerous studies involving systematic
surveys of regional and national water supplies (Cooper and Murchio
1974, Stewart 1976, and Hallenbeck et al. 1977). Elaborate national
surveys of asbestos fiber concentrations in drinking water were
carried out in the late 1970's in the USA (Millette et al. 1979),
in Canada (Health and Welfare Canada 1979, Chatfield and Dillon
1979), in the UK (Commins 1979), in Germany (Meyer 1982, Spurny and
Schormann 1983) and many other places. Based on these surveys it
became evident that most water supplies contain asbestos fibers in
the order of $10^5 - 10^6$ fibers/liter. A relatively small percentage
of all supplies was found to have concentrations greater than 10^7 fibers/liter (USA = 10%, Canada = 5%) and in all cases were
related to areas with naturally occurring asbestos rich bedrock or
water supplies influenced by asbestos mining activities. A brief
summary of the highest asbestos fiber concentration in water
supplies in North America is provided in Table 8 and an indication
of typical fiber size and aspect ratios is given in Table 9.

TABLE 8

Locations with high asbestos fiber concentrations in drinking water in North America

Location	References	Chrysotile Concentration Fiber/Liter
Thetford, Quebec	1	1.7×10^8
Baie Verte, Newfoundland	2	3.2×10^8
Thompson, Manitoba	2	1.9×10^8
Disraeli, Quebec	2	$2.8\text{-}4.1 \times 10^8$
Whitehorse, Yukon Territory	2	2.7×10^8
Santa Fe, New Mexico	3	1.9×10^9 *
Socorro, New Mexico	3	$1.2\text{-}2.1 \times 10^9$ *
Duluth, Minnesota	4	up to 1.5×10^8
Everett, Washington	5	$2.3\text{-}3.8 \times 10^8$
San Francisco Area	6	up to 1.8×10^8
St. Croix, V.I.	7	5.4×10^8 **

* Groundwater
** From Asbestos Tile Roof

References:	1 =	Cunningham & Pontefract (1971)
	2 =	Health & Welfare Canada (1979)
	3 =	Oliver & Murr (1977)
	4 =	Brown et al. (1976)
	5 =	Boatman et al.(1983)
	6 =	Kanarek et al. (1980)
	7 =	Millette et al. (1980a)

TABLE 9

Fiber dimensions and aspect ratios of some asbestos rich drinking water supplies in North America

Source	Average Fiber Size in um		Aspect Ratio	Reference
	length	width		
Chrysotile:				
Silver Lake, NY	0.55	0.08	7:1	Maresca et al.(1984)
San Francisco Bay	1.3	0.04	39:1	Millette et al.(1979)
Tolt River, Wash.	0.8	0.03	25:1	Millette et al.(1980a)
Woodstock, NY	2.2	0.08	27:1	Webber et al. (1988)
Amphibole:				
Lake Superior, Mn.	1.5	0.18	11:1	Langer et al.(1979)

Most water supplies with high asbestos fiber levels are located in areas that have geological formations rich in asbestos fibers.

Probably one of the most affected areas is in northern California where asbestos bearing rocks are abundant. The problem is enhanced by the fact that California is a water deficient state and water distribution systems for drinking water and irrigation are very extensive and interconnected. Once asbestos rich sediment gets into the system the fibers are distributed over large areas, thus affecting drinking water quality for a large segment of population (Hayward 1984, Jones and McGuire 1987). Other areas of North America with high fiber concentrations in drinking water are the Puget Sound area of Washington State, part of the Appalation Mountains in New York, New Jersey, and Maryland, the Gaspai Peninsula in Quebec, the Duluth area in Minnesota, and sections of New Mexico where groundwater was found to contain high fiber concentrations (Oliver and Murr 1977).

Mining and industrial activities have an additional influence on fiber concentrations since they involve the disturbance of large quantities of rock and surface materials. Given the ease with which fibers are spread by wind and water the potential of contaminating water supplies considerable distances away from asbestos sources is significant. Water supplies in Quebec and in Minnesota certainly are an indication of such effects. Finally, the widespread use of asbestos materials in construction and transportation makes the determination of asbestos sources in water supply even more complex. Nevertheless, it appears that natural asbestos sources are more widespread and thus have a greater impact on background levels of asbestos fibers in drinking water, but concentrated mining, industrial use and construction activities often contribute to raising the fiber concentrations to very elevated levels (Peterson 1978, Lawrence and Zimmermann 1977, Bacon et al. 1986).

The use of asbestos cement pipes to transport water is an additional source of fibers in the drinking water. It has been estimated that by 1974 more than 2.4 million kilometres of asbestos cement pipes (A/C pipes) were transporting water world wide (Toft et al. 1984). More recent estimates suggest that over 640,000 km of A/C pipes exist in the USA (Sullivan 1986), and some 320,000 km in the UK (Millette et al. 1981b). Many studies have been carried out to determine how many fibers are introduced into the water by corrosion of A/C pipes. This has led to considerable confusion because in some studies concentrations increased after the water was transported through the A/C pipe network while in other cases the fiber concentrations decreased after transport.

The studies by Oliver & Murr (1977) in New Mexico, Hallenbeck et al. (1978) in Illinois and Millette et al. (1981b) in other parts of the USA found no increase in fiber content before or after flow through A/C pipes. In contrast in Winnipeg, Manitoba, the fiber concentration increased progressively from the water input source to the faucet output (Toft et al. 1981). Clark et al. (1980), Buelow et al. (1980) and Millette et al. (1981b) showed that under specific conditions of asbestos corrosion cement can release asbestos fibers into the drinking water.

The water aggressiveness index proposed by the American Water Works Association Standard Committee has been used as a measure to determine the corrosion potential and hence the ability to release fibers into the water from asbestos cement. The index is derived from the Langlier Calcium Carbonate Index and is defined as:

$$A.I. = pH + Log_{10} (A \times H)$$

where pH is the standard for measuring hydrogen ion concentration,

A is total alkalinity in mg/l as $CaCO_3$ and

H is calcium hardness in mg/l as $CaCO_3$.

According to Millette et al. (1981b) values greater than 12 are generally considered non-aggressive and approximately 31.5% of all water supplies in the USA have values above that level. Highly aggressive water with indices <10 are present in 16.5% of all US water supplies and according to the same authors as many as 40 million people may be exposed to asbestos fibers in drinking water as a result of corrosion of asbestos cement products. However, as shown by Buelow et al. (1980) and Millette and Kinman (1984) the use of the aggressiveness index alone often proves to be insufficient to predict the actual behaviour of A/C pipes. In the absence of calcium carbonates, iron, zinc and manganese can combine with other residue to form protective coatings which limit the leaching of calcium and asbestos fibers from the pipe matrix.

Pipe deterioration in the presence of acid-sulfate rich drainage water from mine tailings containing pyrite was found to be small in a survey in the UK (Lea 1970). In this case the deposition of hydrated ferric oxide inside the pipes formed a protective layer. In the Middle East where sulfate salts are widespread in the soils and water supplies the impact on the pipes has been more significant (Matti and Al-Adeeb 1985). According to Cosette et al. (1986) protective coatings with chlorinated rubber have been developed in some areas where asbestos cement is exposed to aggressive and brackish water. The relationship between protective coating by iron, the aggressiveness of the water, and asbestos fiber release from the A/C pipes is fairly complex and needs further attention (Millette and Kinman 1984).

Another important source of fiber release is tapping. Connecting new pipes to existing A/C pipes is largely responsible for the sporadic introduction of fibers into the water. Some of these activities can result in drastic increases in fiber concentrations particularly in old A/C systems (Webber et al. 1988). Filtration of asbestos rich water is possible and according to Peterson et al. (1980), McMillan et al. (1977) and Hunsinger and Roberts (1980) coagulation and filtration treatments remove fibers efficiently.

As a point of interest, asbestos concentrations of $10^6 - 10^7$ fibers/liter have been reported in beverages such as beer, wine and soft drinks (Biles and Emerson 1968, Cunningham and Pontefract 1971, Gaudichet 1978). The presence of asbestos fibers in such beverages is likely due to the use of water which already contains asbestos fibers, or to airborne contamination, or to input during the bottling and processing where the beverages are commonly filtered through asbestos filters.

Overall it is clear that the major contribution of asbestos fibers to drinking water comes from natural sources. Point sources such as mine operations, quarrying and the use of asbestos material for road construction are the major sources of locally high asbestos fiber concentrations. Asbestos cement pipes and roofing materials can influence the concentrations under conditions of aggressive water and in the absence of protective coating that can occur by natural processes. Finally, the input of fibers via air, rain and snow may be significant in different locations and will be discussed below.

3.2.2 Asbestos Fibers in Rain and Snow

Given the very small size of asbestos fibers such material is readily transported by wind. To differentiate between the natural versus man-made contribution of asbestos in the atmosphere is an exercise in futility except to say that on a world scale natural sources clearly contribute far more fibers to the atmosphere than man-made sources.

Under special circumstances it might be possible to differentiate between industrially processed and natural asbestos fibers in some of the water samples by using dark field electron microscopy techniques (Seshan 1978). Chrysotile fibers tend to have micro-crystalline fiber deformation as a result of industrial milling and mixing processes. Natural unprocessed fibers do not appear to have such deformations. This, however, assumes that no weathering or alteration takes place when the fibers enter the hydrological system. As will be shown in the next section such changes are frequent and the success of separating fiber sources on the basis of micro-crystalline morphology is likely limited.

It is also very difficult to determine how much influence man's activities have on the mobility of fibers in areas where asbestos bearing strata are near the earth's surface. The topic of airborne fibers will be discussed in Chapter 6, however it should be mentioned here that airborne asbestos fibers are collected by rain and snow and returned to the surface via the hydrological network. Few studies have been made to determine the asbestos fiber levels in rain and snow. Hallenbeck et al. (1977) and Hesse et al. (1977) found concentrations of $10^5 - 10^6$ fibers/liter in rainwater in the Chicago area and levels up to 2×10^7 fibers/liter were measured by Bacon et al. (1986) in Quebec. This certainly suggests that atmospheric transport of fibers might be significant in a regional context. Cunningham and Pontefract (1971) also reported detectable fiber levels in snow samples. More measurements are needed to determine the relative contribution of asbestos fibers via the airborne collection process, but according to Cossette et al. (1986) data from ice drill cores collected during the international geophysical year indicate that distribution of global asbestos fibers has been relatively constant, implying that industrial sources have had little impact on the global redistribution of asbestos. However, this is in contrast to findings by Langer (in Selikoff and Lee 1979) who reported higher concentrations in Greenland ice dating from 1920 than older ice.

3.2.3 Asbestos Fibers in Lake and Streams

The three principle sources of asbestos fiber introduction into streams and lakes are: 1. Asbestos bearing rocks and surficial materials in the watershed, 2. Pollution due to effluent from mining and industrial activity, and 3. Atmospheric input. The first source is likely the most widespread affecting many watersheds in the world. The input of asbestos fibers from mining and industrial activity is more localized but can be very significant and atmospheric input can be influenced by both of the other sources.

In the 1970's much attention was given to fiber levels in Lake Superior. Several communities along the lake reported high levels of asbestos fibers in drinking water and the dominantly amphibole fibers were traced to mining activities at Silver Bay in Minnesota. Since 1957 an iron ore mining company has been dumping taconite tailings containing amphibole fibers into Lake Superior. The lake serves as a water supply for a number of communities along the lake and concerns were expressed when Cook et al. (1974) reported asbestos levels in the Duluth water supply ranging from 10^{7}-10^{9} fibers/liter. Other lake samples analyzed by Durham and Pang (1976) showed levels of 0.1×10^{6} - 8.7×10^{7} fibers/liter. The great variability in asbestos fibers observed in the lake was attributed to seasonal and meteorological variations and the water circulation pattern in the lake. Durham and Pang (1975, 1976) noted a decrease in fiber concentrations between June and November, and Fairless (1977) showed decreasing concentrations in a counterclockwise direction from the mine effluent across the lake. To determine the fiber source drill core samples of bottom sediments were analyzed. Sediments deposited prior to 1950 had only traces of amphibole asbestos while sediments from the mid 1960's showed increasing amounts of cummingtonite-grunerite (amosite) asbestos, a mineral dominant in the Silver Bay mining operation (Great Lake Advisory Board 1975). Based on the work by Cunningham and Pontefract (1971), Kay (1973, 1974) and the above mentioned sources it became evident that asbestos fiber concentrations in surface waters in the 10^{5} - 10^{7} fiber/liter range are not at all unusual.

Stream contact with asbestos bearing rocks is the cause of high fiber concentrations in groundwater in New Mexico (Oliver and Murr 1977). The water which flows through serpentine rich rock formations was found to contain up to 10^{9} fibers/liter. A summary of typical asbestos fiber concentrations in rivers and lakes affected by asbestos fibers is provided in Table 10. It can thus be concluded that mineral fibers are a frequent component of the particulate fraction in streams and lakes, but it should be pointed out that not all fibers in streams and lakes are asbestos fibers

(Puffer et al. 1987, Millette et al. 1987). Careful electron microscopic analysis is needed to differentiate between asbestos fibers and attalpulgite, rutile and other fibrous materials.

TABLE 10

Asbestos fiber concentrations in selective rivers and lakes

Water Source	Reference	Asbestos Fiber Concentration Fibers/Liter
Great Lakes:		
Lake Superior	Cook et al. (1976) & Durham & Pang (1975)	$10^5 - 10^9$
Lake Ontario	Kay (1973)	10^6
Lake Huron	Durham & Pang (1976)	$10^6 - 10^7$
Lake Michigan	Cunningham & Pontefract (1971)	$10^6 - 10^8$
Lakes in California:		
Oroville	Stewart et al. (1976)	10^6
Reservoirs	Bales et al, (1976)	$10^{10} - 10^{11}$
Marine County Lakes	Cooper & Murchio (1974)	10^8
Silverwood	McGuire et al. (1982)	$10^8 - 10^9$
Silver Lake	Maresca et al. (1984)	$10^6 - 10^7$
Lakes in Quebec:		
Memphremagog, Quebec	Bacon et al. (1986)	$10^6 - 10^8$
Rivers in Canada:		
Ottawa, Ont.	Cunningham & Pontefract (1971)	10^6
Sumas, B.C.& Wash.	Schreier & Taylor (1981) & Schreier (1987)	$10^6 - 10^{13}$
Fraser, B.C.	Schreier & Taylor (1980)	$10^6 - 10^9$
Yukon, Yukon Terr.	Schreier & Taylor (1980)	$10^7 - 10^8$
Becancour, Quebec	Monaro et al. (1983)	$10^8 - 10^{10}$
Missisquoi, Quebec	Bacon et al. (1986)	$10^7 - 10^8$
Rivers in the USA:		
Trinity River, Cal.	Stewart et al. (1976)	10^7
Beaverhead River	Stewart et al. (1976)	10^7
Sultan River, Wash.	Polissar et al. (1982)	$10^7 - 10^8$
Sacramento, Cal.	Bales et al. (1976)	$10^8 - 10^{11}$
Sacram. San Joaquin	Hayward (1984)	$10^7 - 10^9$
Calif. Aqueduct	Jones & McGuire (1987)	$10^7 - 10^{10}$
Klamath, Cal.	Hayward (1984)	$10^{10} - 10^{11}$

A comparison between asbestos fibers from natural and industrial input sources is provided in Table 11. It shows that fiber concentrations in the natural environment are often equivalent to and in cases greater than industrial point sources. Stream systems with naturally high asbestos fiber sources can be used to evaluate the impact of such material on water quality and the aquatic biota, a subject which will be discussed later in this chapter.

TABLE 11

Comparison of asbestos fiber concentrations between natural water and water influenced by industrial effluent

Source	Natural Water Unaffected by Asbestos Rich Geology	Natural Water Affected by Asbestos Rich Geology	Water Affected by Effluent from Asbestos Industry and Mining Activity
Lakes	$< 10^6$ (1)	$10^6 - 10^8$ (2,3) $10^7 - 10^9$ (19)	$10^6 - 10^9$ (4.5)
Rivers & Effluents	$< 10^6$ (4)	$10^6 - 10^{13}$ (6) $10^7 - 10^8$ (7) $10^6 - 10^9$ (8,11,20) $10^8 - 10^{10}$ (9) $10^7 - 10^{11}$ (21)	$10^6 - 10^7$ (10) $10^7 - 10^{10}$ (10) $10^8 - 10^{12}$ (10) $10^9 - 10^{11}$ (16)
Water Supplies	$< 10^6$ (11) $< 10^6$ (12)	$10^7 - 10^8$ (17) $10^6 - 10^8$ (13,14)	$10^6 - 10^7$ (11,15) $10^6 - 10^8$ (7,12,18)

References:
(1) Kay (1973)
(2) Stewart et al. (1976)
(3) Cooper & Murchio (1974)
(4) Cunningham & Pontefract (1971)
(5) Hallenbeck et al. (1977)
(6) Schreier & Taylor (1981)
(7) Bacon et al. (1986)
(8) Schreier & Taylor (1980)
(9) Monaro et al. (1983)
(10) Stewart et al.(1976)
(11) Millette et al. (1979)
(12) Wigle (1977)
(33) Kanarek et al. (1980)
(14) Polissar et al. (1982)
(15) Millette et al. (1983)
(16) Lawrence & Zimmermann (1977)
(17) Toft et al. (1984)
(18) Webber et al. (1988)
(19) McGuire et al. (1982)
(20) Hayward (1984)
(21) Jones & McGuire (1987)

3.3 Asbestos Fiber Transport and Deterioration in Water

Most asbestos fibers have very small dimensions and are therefore highly mobile. Minor turbulence in water can result in fiber suspension and transport over considerable distances. In the process the fibers can be altered physically and chemically and this may impact upon water quality and the environment.

3.3.1 Sediment Transport

Because there are large seasonal fluctuations in stream discharge and lake circulation patterns it is obvious that the suspended fiber concentration is greatly influenced by discharge and flow patterns. Schreier and Taylor (1980) reported significant seasonal fluctuations in asbestos fiber concentrations in a stream unaffected by asbestos mining and industrial activity. Similar flow related variations were reported by Cook et al. (1974), Bacon et al. (1986), Monaro et al. (1983) and Hayward (1984). Given the fact that the majority of fibers are smaller than 1 um in length they remain in suspension for greater lengths of time than other particles which are usually larger and have different shapes and densities. The shape of the particles is of particular interest since the settling pattern of fibers is likely very different from those of spherical particles which are more common in sediments. If small fibrils do not become trapped in bundles they are likely to remain suspended in water for long periods of time and are therefore distributed over larger distances. Coagulation models discussed by Bales (1985) suggest that about one in every 1000 collisions between submicron fibers and larger particles result in aggregation. The same author presents data to suggest that in a reservoir the asbestos fibers are removed by settling at the order of one magnitude per year. The surface charges of the fibrous particles may also play a significant role in coagulation and sedimentation (Bales 1985). As a result of seasonal flow fluctuations, differences in fiber charges and fiber interactions, and the difference in settling rates of the particles, the task of arriving at actual fiber concentrations in streams and lakes is complex and labour intensive.

Information about the settling patterns of amphibole fibers has been reported by Martilla (1979) in Wabush Lake, Labrador and by Fairless (1977) in Lake Superior. In both cases mine tailings from iron ore mining activity are thought to be responsible for introducing asbestos fibers into the water.

The use of turbidity as an indicator of fiber concentration has resulted in mixed success. Severson et al. (1981) found that turbidity might be used as a surrogate measure of asbestos fiber concentrations in raw water of one watershed, but Lawrence and Zimmermann (1976), McGuire et al. (1982) and Hayward (1984) were unable to find a systematic relationship between turbidity and asbestos fiber counts in their studies. Chatfield et al. (1984) measured turbidity in combination with magnetic separation and claimed some success in measuring high iron amphibole fiber levels in laboratory samples.

In most rivers there are large fluctuations in the water table between high and low discharge levels and significant amounts of asbestos rich sediments can be exposed in depositional areas. Since vegetation cover is limited asbestos fibers can readily be picked up by wind thus posing a potential health hazard to the local population. Also massive amounts of asbestos rich sediments are deposited in rivers, canals and reservoirs and the clogging of such water systems and subsequent removal of sediment is an ever increasing problem. The most notorious case is in the central irrigation and water supply system of central California. Large quantities of asbestos rich sediments are introduced into the canal system every spring and the system has to be dredged frequently to reduce the asbestos fiber concentrations and to prevent the clogging of channels. The extent of the problem and the dredging work have been described by McGuire et al. (1982), Hayward (1984) and Jones and McGuire (1987) and will be further discussed in Chapter 6.

3.3.2 Physical Deterioration of Fibers

Because of physical wear, saltation of particles and abrasion it is likely that asbestos fibers break into smaller components when transported in turbulent streams. Due to physical wear the fiber bundles are likely to disintegrate into fibrils. The relationship between fiber and fibril length was demonstrated by Harington et al. (1975) and has been shown in Figure 2. This is of interest since thin and long fibers are considered more carcinogenic than shorter and fatter fibers. Decreases in fiber length as a result of transport and physical wear has been shown in the laboratory by Spurny (1981). This is however more difficult to document in river systems where particles settle at different rates depending on velocity. The larger particles settle under higher discharge velocity, while the smaller particles settle in calmer water. It is thus difficult to say that asbestos fibers are smaller downstream due to physical wear when settling patterns, turbulence and velocity determine the size distribution in the sediments. Evidence to this end was provided by Schreier and Taylor (1981). Stokes' law states that the settling rate of a falling particle is directly proportional to its size squared. This applies to spherical particles under the assumption that particles do not interact with each other during the fall. None of these conditions apply since asbestos fibers are not spherical, they do interact by bundling and coagulation, and as shown under the mineralogy section asbestos fibers have different charges in aqueous media (Seshan 1983). Therefore attraction or dispersals between particles is likely to happen, greatly complicating the behaviour of fibers in stream and lake systems. In addition the measurements and prediction of fiber size are complex (Siegrist and Wylie 1980) given the abnormality of many sample distributions, the limits of analytical resolution, sample preparation and contamination problems. Nevertheless it can be stated that asbestos fibers are generally smaller in streamwater than those used in industry, since the latter has a much more limited distribution in the world and has undergone less secondary modification.

3.3.3 Chemical Deterioration of Fibers

As shown in laboratory experiments asbestos fibers have widely different resistance in aqueous, acidic and alkaline environments. Once the asbestos fibers are introduced into the water supply leaching occurs and this might alter the surface characteristics of the fiber and its behaviour.

Chrysotile fibers are very stable in alkaline water but the leaching of magnesium from the fiber structure can readily take place under acidic conditions. Since the majority of river systems are acidic it is likely that this process takes place in nature. According to Choi and Smith (1972) the rate of magnesium and OH dissolution is controlled by diffusion from the surface into the water. However, Bales and Morgan (1985a) showed that the rate of magnesium over silica release is 2:1 in the pH range 7-9 and, given the slow rate, suggest that chemical reaction rather than diffusion is rate limiting. As shown by Verlinden et al. (1984) and Bales and Morgan (1985a) organic components and oxalates affect the rate of magnesium release in the early stages of leaching. In addition, chrysotile fibers freshly suspended in water below pH 9.8 have positive surface charges but change with time to negative charges due to rapid adsorption of organic particles eventually covering the fiber surface completely (Bales and Morgan 1985b, Seshan 1983). It should also be noted that the leaching of magnesium is not constant along the fiber axis (Spurny 1983) and this further complicates the analysis.

Aquatic systems also show differences in Mg/Ca relationships. The Mg/Ca ratios are usually a good indicator of the influence of serpentinitic rock formations on the stream system. An excellent example of this effect is the water quality in the Sumas river in NW Washington and British Columbia. As shown by Schreier and Taylor (1981) and Schreier (1987) a landslide in the headwaters of the river has exposed an asbestos rich serpentinitic bedrock formation. Large quantities of asbestos fibers are introduced into the stream system at the landslide source area in the headwaters and the magnesium concentrations in water and sediments are largest at that point in the stream system and progressively decrease with increasing distance from the source area (Figures 7 and 8).

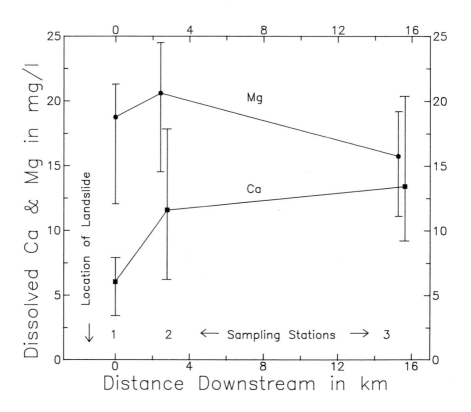

Fig. 7. Calcium and magnesium concentrations in stream water with increasing distance downstream from the asbestos fiber input source in the Sumas River in B.C.

As shown in Figure 7 and 8, the opposite effect is seen with calcium. The concentrations progressively increase with increasing distance downstream. In the majority of streams calcium values in both water and sediments are generally larger than magnesium levels.

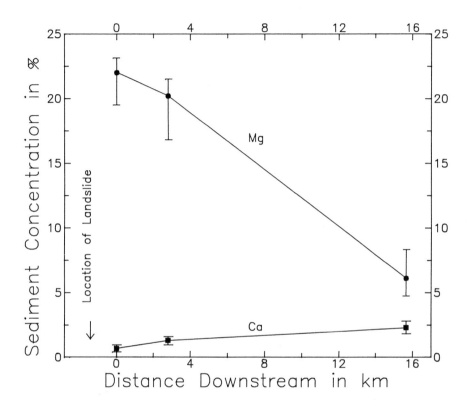

Fig. 8. Calcium and magnesium concentrations in sediments with increasing distance downstream from the asbestos fiber input source in the Sumas River in B.C.

The Ca/Mg ratio could thus be used as a crude index in determining the impact of chrysotile asbestos on stream water and it could be postulated that where calcium values exceed magnesium values the chemical effect of asbestos is no longer important. Monaro et al. (1983) also used magnesium concentrations to indicate the impact of asbestos rich mine effluences in Quebec rivers. In fact magnesium concentrations in impacted rivers are generally correlated with fiber concentration. As illustrated in Figure 9 we found a good relationship between asbestos fiber concentration and Mg/Fe ratio in stream water in the Sumas River in British Columbia. A note of caution is in place. Such relationships are very basin specific and given the analytical problems in determining accurate fiber counts, the great variability in asbestos bearing bedrock in different basins, and the many factors that can influence magnesium and iron levels in streams, such predictive relationships only give a crude idea of chrysotile fiber concentrations (Monaro et al. 1983, Schreier 1987).

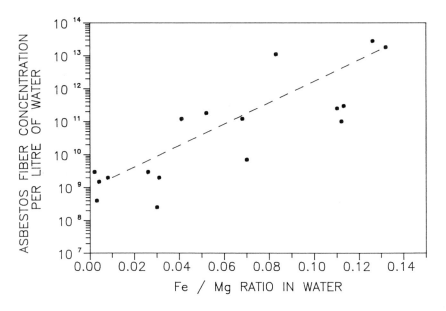

Fig. 9. Relationship between asbestos fiber concentration and Fe/Mg ratio in stream water in the Sumas River basin in B.C.

Another important effect of asbestos fiber impact on water quality relates to the presence of trace metals. As mentioned in the mineralogical sections large quantities of trace metals are present in asbestos fibers and once exposed to weathering these trace metals are released into the water system. Nickel, chromium, cobalt and to a lesser extent manganese concentrations increase significantly in the water and sediments when high levels of asbestos fibers are present in the stream system. Similar to the magnesium example in the Sumas river, the concentrations of nickel, chromium and cobalt were highest at the point input source of asbestos and steadily declined with increasing distance downstream and away from the source area (Figure 10). Since these trace metals are more unique than magnesium they serve as a better and more easily measurable indicator of asbestos fibers in stream water.

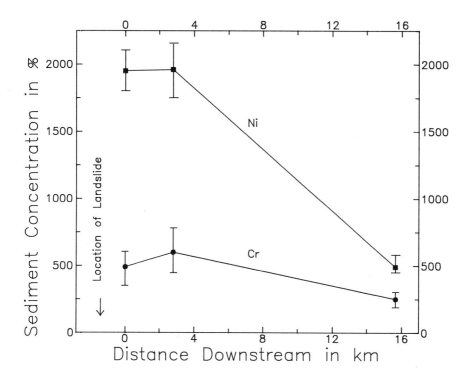

Fig. 10. Trace metal distribution in stream sediments with increasing distance downstream from the asbestos fiber input source in the Sumas River in B.C.

The presence of organic acids in stream water can likely result in rapid alterations of chrysotile fibers and it appears that although the fiber structure remains after magnesium is removed the structure becomes gelatinous or colloidal (Spurny 1982, Seshan 1983, Bellmann et al. 1986) and likely disintegrates in turbulent conditions, interacting to form secondary clay minerals such as smectites (Schreier et al. 1987a). The leaching is further enhanced in streams because of fiber dispersion and possible defibrillization which, as shown in the laboratory, increases the rate of leaching significantly (Monkman 1971, Atkinson and Rickards

1971, Harris and Grimshaw 1971).

The leaching is of interest in terms of waste disposal of asbestos fibers and will be discussed in a later chapter. Also, the fact that acid leached chrysotile fibers have been shown to be less carcinogenic in animal studies (e.g. Morgan et al. 1977, Monchaux et al. 1981) is of relevance since some leaching will take place in rivers due to natural acidity. This might be an explanation of why health hazards from ingesting chrysotile fibers in water are now considered minimal. More details on this topic can be found in section 3.5.

Amphibole fibers generally have less chemical impact on water quality because they are mostly acid resistant hence do not leach as rapidly as chrysotile fibers in the generally acidic drainage water. Only laboratory leaching studies are available and some of these have shown that tremolite fibers, which have recently been suspected to be the major cause of cancer by inhalation (Wagner 1986), are very resistant to leaching even in the presence of oxalates which are a strong chelating agent (Mast and Drever 1987). Similarly, crocidolite (Bellmann et al. 1986) and to a lesser extent amosite (Ralston and Kitchener 1975) proved to be very resistant to acid leaching conditions but all asbestos fibers were found to have chemically reactive surface sites (Leight and Wei 1977, Bonneau et al. 1986).

Defibrillization of amphibole asbestos may liberate many unsatisfied sites in the fiber structure and these might react with trace metals and organic compounds in the water (Langer and Nolan 1986).

3.4 Impact of Asbestos Fibers on the Aquatic Biota

Efforts have been made to determine the potential impact of water borne asbestos fibers on human health. In contrast, the effect of fibers on the aquatic biota has largely been ignored (Henson 1985). Yet, it is known that fish develop tumerous growths and it has been suggested that fish be used as cancer testing animals (Stewart 1977). This has a number of attractive advantages over conventional testing methods. The Smithsonian Institute in Washington D.C. maintains a Tumor Registry in Lower Animals and many of the specimen are fish species, but no link between asbestos and fish cancer has yet been reported.

3.4.1 <u>Impact on Fish</u>

Some of the earliest tests determining the effects of asbestos fibers on fish were carried out by Batterman and Cook (1981) who were interested in determining whether fish exposed to amphibole asbestos fibers in Lake Superior accumulate the fibers in the kidney and muscle tissue. They found fibers predominantly in the kidney and to a lesser extent in some of the muscle tissue but concluded that the accumulation of fibers was very low and of no concern for human consumption. Oxberry et al. (1978) examined whether taconite tailings, which are the principle source of amphibole fibers in Lake Superior, had adverse effects on fish in long term experiments. They exposed juvenile fish, trout embryos and fry, amphipods and mysids to acidified taconite tailings and found no adverse effects on the animals.

Evidence of possible adverse effects of asbestos fibers on fish was, however, provided by Belanger et al. (1986c) who found behaviourial and histopathological aberrations and a few tumorous swellings in coho salmon larva when reared in chrysotile rich water at concentrations of 3 million fibers\liter. They suggested that asbestos stressed fish may also be more susceptible to other waterborne pollutants. Mesothelioma has been reported in fish (Herman 1985) but the author made no reference to possible links with asbestos exposure.

As previously mentioned, trace metals have been used as tracers of asbestos fiber pathways in rivers and terrestrial systems (Schreier et al. 1986, 1987b, and Schreier 1987). The same method was used to analyze fish from the Sumas River in British Columbia where chrysotile asbestos fiber concentrations range from 10^7 - 10^{12} chrysotile fibers/liter. The fish showed no evidence of unusual growth but recorded significant levels of nickel and manganese in the muscle tissue of a variety of fish species. The effects of chrysotile asbestos on fish health and evidence of trace metal accumulation are being further tested in trout originating from the Serpentine Lake in British Columbia (Schreier et al. in preparation).

3.4.2 <u>Impact on Other Aquatic Organisms</u>

Henson (1985) reported a decrease in longevity of Daphnia with increasing asbestos fibers and Bellanger et al. (1986a & b, and 1987) noted a significant reduction in siphoning activity and an alteration of growth and reproduction in juvenile Corbicula Fluminea (asiatic clam) when exposed to chrysotile asbestos. The same authors suggest that Corbicula may be a useful biomonitor since it accumulates asbestos fibers. In addidion, these clams have also been reported to accumulate trace metals (Graney et al. 1983) and it would be most useful to determine asbestos fiber, nickel, manganese, chromium and cobalt content in such clams. If a good relationship between fiber and trace metal content can be established the metal content which is much easier to determine could then be used as a surrogate predictor of asbestos fiber exposure.

Finally, Lauth and Schurr (1983 and 1984) determined the effects of chrysotile exposure to planktonic algae (cryptomonas erosa). They found that asbestos fibers accumulate in the algae and proposed that the positively charged chrysotile fibers will attach to planktonic cells, inhibiting effective swimming and removing plankton, which is a major component in the food chain, from the water column.

While the above mentioned studies are suggestive of adverse effects of asbestos fibers on aquatic biota, there are still uncertainties and disagreements and more research is needed so that a more enlightened assessment of possible problems can be made.

3.5 Impact on Human Health

Virtually hundreds of papers have been published on the topic of possible effects of ingested asbestos fibers on drinking water and human health. As reviewed by Millette et al. (1981a) a number of studies showed that occupational exposure to asbestos fibers resulted in a significantly greater incidence of digestive system cancer. Many animal feeding experiments and epidemiological studies were initiated to determine whether or not ingested asbestos fibers lead to higher incidence of peritoneal mesotheliomas and gastrointestinal cancer in the general population.

3.5.1 Animal Feeding Studies

Some animals exposed to asbestos by inhalation and injection were found to develop malignant neoplasms but no evidence of increased cancer incidence was obtained from the many long term animal feeding studies carried out with hamsters (Smith et al. 1980, McConnel et al. 1983a), baboons (Webster 1974) and rats

(Gross et al. 1974, Bonser and Clayson 1967, Wagner et al. 1975, Hilding et al. 1981, Bolton et al. 1982, McConnel et al. 1983b). In almost all cases the researchers were unable to show significant links between either chrysotile or amphibole fiber ingestion and gastrointestinal cancer.

3.5.2 Epidemiological studies

Since extrapolation from animals to humans is always controversial a series of epidemiological studies was undertaken to show possible links between asbestos fibers ingested via drinking water and gastrointestinal cancer in the general population. Millette (1983), Toft et al. (1984), Cossette et al. (1986), Commins (1984) and MacRae (1988) all reviewed some of the major studies carried out in Minnesota, Quebec, California, Washington State, Connecticut, and Florida, where asbestos fiber concentrations in drinking water have been reported to be very high. A summary of the studies with appropriate references is provided in Table 12. No unanimous conclusion could be reached by the many authors of the projects and this is not unexpected since most epidemiological studies are influenced by many different factors. The majority of the findings do however point to the fact that cancer risk from asbestos fibers ingested via drinking water is small (Wigle et al. 1986, DHHS Committee 1987, MacRae 1988).

3.6 Summary of Impact on Health by Waterborne Asbestos Fibers

Most authors have concluded that there is little solid evidence from both animal feeding studies and epidemiological research that exposure to asbestos in drinking water leads to adverse health effects (Toft et al. 1981, Millette Ed. 1983, Commins 1983 & 1985, Cossette et al. 1986, Langer and Nolan 1986, Toft and Meek 1986, DDHS Committee). As noted by Langer and Nolan (1986) if there is a risk it is very small and well below the resolution power of epidemiological studies.

Some important questions remain because there are several cases reviewed by Velema (1987) and Toft et al. (1984) in which the apparent relative risk of stomach and pancreas cancer from fibers ingested via drinking water is elevated. Similarly, some association between ingested asbestos fibers and cancer induction exists as shown by epidemiological research (Kanarek et al. 1980, Polissar et al. 1983, and Toft et al. (1984) and animal feeding studies (McConnel et al. 1984). Unfortunately it is almost impossible to control the many potentially confounding variables inherent in the methodologies of both animal feeding and epidemiological studies. The major shortcomings in this regard are listed by Toft et al. (1984) and MacRae (1988) as follows: Problems with fiber measurements and contamination of samples, confounding problems with dietary intake, possible synergistic interactions, long latency periods, high population mobility, differential exposure to airborne fibers, confounding problems between inhaled and ingested fibers, and possible migration of fibers through tissues and cell walls.

TABLE 12

Characteristics of asbestos fibers in drinking water where major epidemiological studies have been carried out

Area	Type of Asbestos Fibers	Source	Concentration Range in Water (MFL)	Mean Fiber Length um	Years of Popul. Exposure	Relevant References
Duluth, Minn.	Amphibole	Mining	1×10^6 to 3×10^7 1.5 (6% >5)		< 20	1,2,3,4,5,
Quebec,Canada	Chrysotile	Natural & Mining	1×10^6 to 1×10^9 (2% >5)		> 50	6,7,
San Francisco Bay Area, Cal.	Chrysotile	Natural	1×10^5 to 4×10^7 1.4 (2.5%>5)		> 50	8,9,10,11,
Seattle, Wash.	Chrysotile	Natural	4×10^6 to 2×10^8 (<1% >5)		> 50	12,13,5,
Cunnecticut	Chrysotile	Asbestos Cement Pipes	$<10^5$ to 1×10^6 (16% >5)		< 15	14,15,
Escandido, Fl.	Chrysotile	Asbestos	7×10^5 to 3×10^7		> 25	5,16

References:
1 = Mason et al. (1974)   9 = Conforti et al. (1981)
2 = Levy et al. (1976)   10 = Conforti (1983)
3 = Sigurdson (1983)   11 = Tartar et al. (1983)
4 = Sigurdson et al. (1981) 12 = Polissar et al. (1982)
5 = Millette et al. (1980a) 13 = Polissar et al. (1983)
6 = Wigle (1977)   14 = Harrington et al. (1978)
7 = Marsh (1983)   15 = Millette et al. (1983)
8 = Kanarek et al. (1980)

The current consensus amongst scientists is that the potential risks associated with asbestos fiber digestion and digestive system cancer is very small and the sensitivity and resolution power of the currently used epidemiological and animal feeding studies are too small to provide enlightening results. The topic can most appropriately be left by quoting Langer and Nolan (1986): " No measurable effect is not the same as no effect ".

Finally it should also be mentioned that water moves and redistributes asbestos fibers. Once the water evaporates from the depositional environment the asbestos fibers are then exposed to wind action and might pose a health hazard in terms of inhalation. The topic of asbestos-contaminated drinking water and its impact on household air has been addressed by Webber et al. (1988) who present data which show a possible relationship. Similar concerns have also been raised by Schreier et al. (1986) in fluvial situations where asbestos rich sediments were deposited on agricultural land during flooding periods. Because of the unusual biophysical conditions of the material it is difficult to vegetate such sites and the sediments thus become a good source of air pollution. This raises the questions of potential hazards to the rural population by inhalation. Little work has been done on these topics but the potential effects are considered much smaller than in industrial settings since the fiber size is generally smaller after water transport, and chemical modifications of chrysotile fibers by acid leaching in the water are also expected to reduce the hazards. This latter point is significant because studies by Morgan et al. (1977), Monchaux et al. (1981), Jaurand et al. (1984), and Harvey et al. (1984) have shown that the cancer incidence in test animals is sharply reduced if asbestos fibers are leached with acids prior to administration to animals.

CHAPTER 4

ASBESTOS FIBERS IN THE SOIL ENVIRONMENT

4.1 <u>Introduction</u>
    The term "Serpentine soils" is used extensively in botanical
literature to describe soils which harbour unique plant
communities. The vegetation on such soils is often unique and in
contrast with the surroundings vegetation. These soils originate
on ultramafic bedrock that has been altered by metamorphic and
hydrothermal processes, but they can also form on altered
sedimentary rocks such as dolomite. Serpentine rocks do not cover
extensive areas but they have a worldwide distribution and have
been given much attention since early history because plants
respond to these soils in a very peculiar manner. Soil evaluations
have concentrated on determining nutrient deficiencies and excesses
in trace metal but few studies have dealt with soil genesis. Given
the vast interest in asbestos in relation to human health and
industrial activity it is quite surprising how little attention has
been given to the topic of asbestos fibers in these soils.
    Chrysotile asbestos is a constituent of most serpentine soils.
It is reported as being present in the soil parent material but
has so far eluded extensive examination in pedology. What is of
interest are such things as: type of fiber present in the soil,
resistance and alteration of asbestos fibers, the impact of fibers
on soil biology and plant growth, and possible mobility of fibers
from soils to plants and soils to air. Very little has been written
on these topics and an attempt will be made in this chapter to
summarize the available literature and to suggest where pertinent
research efforts should be made.
    The topics to be addressed in this chapter are: asbestos rich
soils, their mineralogy, genesis and evolution, the nutrient and
trace metal problems associated with these soils, their impact on
plant growth and soil biology, methods and techniques to modify
and improve these soils to enhance vegetation growth, and the
impact of such soils on human health.

4.2 <u>Mineralogy, Soil Genesis and Evolution</u>

The term "Serpentinitic Soils" has been used in a very loose context and is generally given to soils developed on altered ultramafic rocks. Ultramafic is a general term describing rocks that consist of more than 70% ferromagnesium minerals. However, not all ultramafic rocks are serpentinites and a large number of other rocks are often included in the ultramafic group due to their somewhat similar composition, and the soils derived from most ultramafic rocks exhibit similar stress on vegetation growth.

4.2.1 <u>Mineralogy of asbestos rich parent materials</u>

In the strictest sense serpentinitic soils are restricted to serpentine bedrock which is a rock made up of antigorite, lizardite and chrysotile minerals, and such accessory minerals as magnetite, chromite, biotite and selected amphiboles. Ultramafic rocks known as dunite, periodotite and pyroxenites are often altered by metamorphic processes and secondary hydrothermal alterations to form serpentinites. Serpentinitic rocks and minerals can also be formed by metamorphic transformation of dolomite but in all cases the host rock undergoes recrystallization and chemical changes. One type of ultramafic rock called the alpine type peridoties-serpentine rock association is sometimes referred to as ophiolites and these rock formations are closely associated with the boundaries of the world's tectonic plates (Coleman 1977).

As shown by Dixon (1977) serpentine minerals are not known to form in soils. They are inherited from the parent material and tend to weather into secondary clay minerals. Given the fact that these soils have unusual composition, many nutrient imbalances, and unique plant communities associated with them, it is surprising how little we know about the genesis and evolution of these soils. Krause (1958), Proctor and Woodell (1975), and Brooks (1987) have reviewed the literature and concluded that such soils have unique conditions and many peculiarities but the variability from site to site is so great that it is difficult to generalize. There are many reasons for the variability. Mineralogical variations, polygenesis, climatic variation, inherent site conditions and time differences are probably the most important.

4.2.2 Weathering and Soil Genesis

As shown in Chapter 2 it is difficult to characterize serpentine and amphibole minerals chemically, given the many isomorphic substitutions present in such minerals. In fact hydrothermally altered rocks are renowned for their extreme variability since the temperature of formation and composition of the formation water are different in each location.

Soils developed from serpentinitic bedrock are often polygenetic which implies that modifications in parent materials and weathering processes have occurred over time. Residual soils which develop in situ are rare in the Northern hemisphere since glaciation has disturbed and redistributed soils and weathered surface materials. Subsequently other materials have been added to the soil by glacial, fluvial, aeolian and colluvial processes. A good example of the latter can be found in the Yalakom River basin in North-Central British Columbia where soils developed on serpentinitic soils have been covered with substantial deposits of volcanic ash. These processes mix and often dilute the soils so that they partially lose their distinctly serpentine characteristics.

Climatic variations also have a great influence on variability. Serpentinitic soils under wet tropical environments are commonly leached to form a lateritic type of soil. Magnesium and silica are preferentially leached from the profile and sesquioxides rich in aluminum and iron usually accumulate. Schellman (1964) and Jones et al. (1982) have shown that gibbsite and goethite are the dominant weathering products in these tropical climates. In more temperate environments the development of a clay enriched (argillic) B-horizon has been reported by Wildman et al. (1968a, 1971) in California and Oregon, Rabenhorst et al. (1982a) in Maryland, and Wilson and Berrow (1978) in Scotland. All of the above authors noted that smectite formation is favoured at some stage of weathering. Paquet et al. (1981) also found smectite in a weathered laterite originating from ultrabasic rocks in West Africa, and Istock and Harward (1982) identified smectite derived from serpentine and/or chlorite in poorly drained sites in Oregon. In addition to smectites, other weathering products such as chlorite, vermiculite and talc have also been reported in serpentine soils in Japan (Kanno et al. 1965), Maryland (Rabenhorst et al. 1982a), France (Decloux et al. 1976), Canada (Schreier et al. 1987a), Oregon (Alexander 1988), New Zealand (Kirkman 1975), and West Africa (Paquet et al. 1981). These clay minerals are likely developed under well drained conditions and are also influenced by the addition of aeolian loess, volcanic ash and colluvial materials from surrounding rocks (Kanno et al. 1965, Sticher et al. 1975, Rabenhorst et al. 1982a). Similar clay minerals were also observed in soils developed from amphibole rich (actinolite-tremolite) parent materials in Maryland. Rabenhorst et al. (1982b) identified interstratified chlorite/vermiculite as major weathering products. The formation of nickel silicate as a weathering product of ultramafic rock has also been reported by Schellmann (1982).

A possible weathering sequence for serpentinitic bedrock is provided in Figure 11 and is largely dependent on the purity of the rock material, the intensity of weathering, pH, leaching environment and drainage. According to mineral stability sequences olivine, pyroxene and serpentine minerals are some of the most weatherable rock components, but as shown by Proctor and Woodell (1975) the rate of serpentine weathering is extremely variable. Schott and Berner (1985) showed that olivine and pyroxene weathering might be controlled by surface reactions rather than solution transport mechanisms. These iron rich minerals formed hydrated ferric oxide layers on the outside and an iron-magnesium silicate beneath. At pH levels above 6 these protective layers continue to grow and likely inhibit dissolution of primary minerals. A similar mechanism might apply to the acicular amphiboles.

The carbon dioxide content and the presence of organic acids in the weathering solution has a significant impact on the rate of magnesium and silica removal from the serpentine rock and asbestos fiber structure (Wildman et al. (1968b). At high concentrations of carbon dioxide, in the presence of organic acids, and under free drainage magnesium is leached preferentially over silica, iron, and aluminum. The loss of magnesium and the presence of amorphous silica in the upper horizon of a chrysotile rich soil in Oregon described by Istock and Harward (1982) is a good example of this process. This is in close agreement with findings from laboratory leaching studies by Atkinson (1973), Spurny (1982), Seshan (1983), and Bellmann et al. (1986), who reported magnesium losses from chrysotile fibers and a gel or amorphous silica fiber residue. Lichen have also been reported to be a very effective agent of

72

weathering of serpentine rocks by Wilson et al. (1981) and Ritter-Studnicka and Klement (1968).

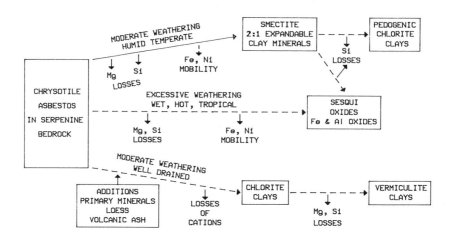

Fig. 11. Weathering sequence of asbestos rich serpentinitic soils. (after Rabenhorst et al. 1982, Schreier et al. 1987a).

According to soil genesis most soils derived from serpentine lose magnesium and silica preferentially over iron and aluminum but the role of organic matter and drainage is critical. Under organic rich, well drained conditions and intensive water movement iron and aluminum can be mobilized by complexing with organic matter. Under such conditions the formation of podzols (spodosols) is a possible end member soil. However, Juchler and Sticher (1985) found only Inceptisols in typical podzol environments on residual serpentine in Switzerland while Kirkman (1975) and Lee and Hewitt (1981) reported moderately leached soils approaching Spodosols in base rich material in New Zealand. De Kimpe et al. (1973) documented how serpentine rich materials added by glacial mixing and aeolian processes can retard leaching and podzolisation. In low organic matter environments sesquioxides remain after the more mobile components such as base cations and silica have been leached from the profile. Lateritic type soils such as Oxisols and Ultisols are considered the end member soil (Alexander 1988). Depending on the degree and length of weathering serpentinitic soils can be classified into many different soil orders. In the more temperate climates Entisols, Inceptisols, Alfisols, Mollisols, and Vertisols have developed on serpentine parent material (Alexander et al. 1985 and Rabenhorst et al. 1982a). Oxisols and Ultisols are more common on serpentines in tropical environments (Sahu 1981, Ogura et al. 1981, Alexander 1988).

4.3 Nutrient Problems

Native vegetation growth on asbestos rich serpentinitic soils has been the subject of much curiosity by botanists. These soils are known to provide unfavourable environments for plant growth, harbouring unique plant communities, exerting stress on some plants and giving rise to limited vegetation cover and low species diversity. In spite of the massive literature on the subject (Proctor and Woodell 1975, Brooks 1987) it is still unclear what the ultimate causes of these problems are. There is no one single factor that is responsible for the infertility of such soils and the many major and minor nutrients present in unusual quantities in these soils are all likely contributors to the problem. What is of particular interest is that the highly weathered serpentine soils are usually more fertile than poorly weathered soils and this is in contrast to most other soils in more humid environments where nutrients are often leached from the soil with increasing weathering (Alexander 1988). This would suggest that some components of serpentine are toxic and with increased leaching the toxicity declines due to conversion and removal of toxic components by water and weathering.

4.3.1 Calcium and Magnesium

With the exception of soils developed on dolomite, some soils derived from ultrabasic rocks and soils affected by seawater salts, the total and exchangeable calcium values in soils are usually larger than magnesium values. All serpentinitic soils show the exact opposite trend. In fact the difference is so striking that Rabenhorst and Foss (1981) use the Ca/Mg ratio successfully in soil surveys to distinguish between serpentinitic and non-serpentinitic soils. In serpentine soils this ratio is usually smaller than 1.0 while most other soils show a ratio larger than 1.0 (Kruckeberg 1969, Alexander et al. 1985).

Both are essential macro-nutrients which play a most important role in plant growth, they appear to have a synergistic effect on plant uptake of other nutrients, they reduce the toxic effects by some metals, and influence soil pH and base saturation. Usually the calcium content in serpentine soils is very low because it is not present in the serpentine mineralogy. Many authors have suggested that calcium deficiency plays an important role in determining vegetation growth on serpentine but the response to calcium additions has not always been favourable. Main (1981) and Walker et al. (1955) showed decreasing growth with increasing calcium levels and Proctor (1971) and Meyer (1980) showed positive responses to growth by Ca additions. There are several explanations for this contradiction. Different plant species respond very differently to soil conditions, the interactions between calcium, other nutrients and metal toxins appear significant, and the form in which calcium is added might be important. Meyer (1980), for example, found good plant response when calcium was applied in the form of gypsum, while lime applications did not measurably improve plant growth. It appears that serpentine endemic plants are better adapted to survive and utilize low levels of calcium than most other plants, and additions of calcium might make conditions more tolerable for growing non-endemic plants on serpentine soils, but are not necessary for the endemics (Brooks 1987).

The addition of calcium increases the ability of some plants to utilize soil phosphorus but again this is a very species specific response. Probably the most important function of calcium and magnesium is to protect the plant from toxic effects of metals in the soil (Proctor and Woodell 1975, Robertson 1985).

Calcium appears to be a key factor in influencing serpentine infertility and responses to calcium additions are evident but not consistent. As mentioned by Proctor and Woodell (1975) there is no convincing evidence to suggest that calcium deficiency is responsible for the poor plant response. What is also interesting is that many non-serpentine soils with calcium values lower than the serpentines do not show the type of vegetation selection and stress found on the serpentine soils.

In contrast to calcium, magnesium is most often thought to be present in toxic concentrations in serpentine soils. Concentrations of more than 20% magnesium are not uncommon and reference to magnesium toxicity have been made by Proctor (1970 and 1971) and Kinzel (1982). Many serpentine species have adapted to high magnesium conditions (Walker et al 1955, Madhok and Walker 1969), but there appears to be an antagonistic effect between magnesium and such other nutrients as boron, iron, manganese, potassium, and phosphorous. According to Brooks and Yang (1984) high magnesium levels are not only phytotoxic to some plants but appear to inhibit the uptake of some of these other nutrients.

Much has been written about the unfavourable Ca/Mg ratio in serpentine soils and its role in plant nutrition. As summarized by Brooks (1987) there are drastic differences between total and exchangeable Ca/Mg ratios in serpentine versus non-serpentine soils. The exchangeable ratios in serpentine are often more than one order of magnitude smaller than the Ca/Mg elemental ratios (Shewry and Peterson 1975). This implies that magnesium availability is significantly lower than calcium and suggests that the toxic effect of high magnesium levels is somewhat modified by the significantly lower solubility and availability of magnesium. This was amply documented by Metson and Gibson (1977) who compared the exchangeable and reserve magnesium levels with the total elemental magnesium levels in 32 different soil types in New Zealand. Brooks (1987) also pointed out that the availability of calcium and magnesium might be further inhibited in the presence of oxalates, which form very low solubility products (1.8×10^{-9} and 8.6×10^{-5} respectively). It should also be mentioned that under well drained leaching conditions magnesium is preferentially removed from the serpentine asbestos fibers and this suggests that the Ca/Mg ratio might become closer to 1 with increasing weathering. The possibly adverse effect caused by excess magnesium and calcium deficiency is therefore likely to change over time.

While it is clear that both calcium and magnesium play a significant role in serpentine nutrition our understanding is still very poor and the available data is often contradictory.

4.3.2 Major Plant Nutrients

Besides calcium and magnesium a number of other major plant nutrients are also found in unusual concentrations in serpentine soils. Nitrogen, phosphorus, boron and molybdenum have all been found to be deficient in most asbestos rich soils. The nitrogen deficiency is caused by the lack of organic matter addition which is often insufficient due to poor vegetation cover on such soils. Accumulation of magnesium in roots has been reported by Morgan et al. (1972) to interfere with nitrogen metabolism and protein production. The addition of nitrate-nitrogen to serpentine soils has proven to be undesirable in many fertilization attempts and Proctor and Woodell (1975) and Proctor (1971) suggested that additions of salts increase the magnesium solubility.

Phosphorus is generally less abundant in serpentinitic rocks and phosphorus deficiencies are common (Proctor and Woodell 1975). Iron which is fairly abundant in serpentinitic rocks can combine with phosphorus to form unavailable and insoluble phosphates. Similarly, phosphorus interactions with trace metals are also plausible explanations for available phosphorus deficiencies. Evidence to this end was provided by Soane and Saunder (1959) who observed negative correlations between chromium and phosphorous levels in plants. Bulusu et al. (1978) have shown that soluble phosphate is strongly adsorbed to serpentine minerals in laboratory experiments. In normal soils total phosphorus usually declines with time, but the reverse is the case in serpentine soils. Due to weathering and leaching some of the potentially toxic components are reduced and this leads to greater accumulation of organic matter, hence an increase in organically bonded phosphorous (Lee and Hewitt 1981). The application of mixed NPK fertilizers to alleviate deficiencies has met with mixed success.

Molybdenum and boron deficiency has also been identified as a major nutrient problem in many serpentine soils. Molybdenum is generally mobilized by organic materials in soils with acidic pH (Szilagyi 1967). Serpentine soils rarely have low pH values and organic litter is usually scarce. In addition molybdenum requirements are somewhat higher in commercial plants than in native species, and this might be one of the contributing reasons why reclamation success with commercial species has been poor. A more detailed discussion on this topic is provided in section 5.4.

Boron is not essential to animals but is needed in small quantities for plant growth. The availability of boron is also reduced with lime application and the deficiency in serpentine soil is caused by the low levels of boron in this type of rock and the low availability in alkaline pH. Table 13 provides a comparison between major nutrients in common non-serpentinitic rocks and ultramafic serpentine. The phosphorus, calcium and potassium levels in serpentine are significantly lower than average world conditions and the exact reversal is the case for magnesium.

TABLE 13

Major nutrient elements in asbestos bearing rocks and soils

Source	Ca %	Mg %	Mg/Ca	K %	P ug/g^{-1}
Dunite Rock	0.15	29.7	247	0.05	nd
Chrysotile Fibers	0.11	28.9	300	0.06	nd
SerpentineSoil A	0.08	14.5	181	0.01	186
" " B	0.70	10.2	15	0.58	190
" " C	1.20	20.0	17	nd	600
" " D	0.14	14.0	100	0.05	122
" " E	1.45	7.0	5	0.15	144
" " F	0.70	16.0	23	0.01	108
Non SerpentineSoils					
World average (G)	1.96	0.8	0.4	1.83	800

References:
A = Jaffre (1980) New Caledonia
B = Shewry & Peterson (1976) United Kingdom
C = Schreier et al. (1987) Canada
D = Roberts (1980) Canada
E & F = Slingsby & Brown (1977) United Kingdom
G = Ure & Berrow (1982)

4.3.3 Metal Toxicity

Metals such as iron, cobalt, chromium, nickel and manganese have all been reported to be in very high concentrations in different serpentinitic soils around the world and much attention has been devoted to the topic of metal toxicity. The total elemental content of trace metals in soils does not directly reflect the problems of toxicity to plants because the soluble, available and exchangeable portions of these trace metals are usually low and dependent on pH and redox potential. The subject has been thoroughly reviewed by Proctor and Woodell (1975), Slingsby and Brown (1977), Sasse (1979a), and Brooks (1987). The latter author lists more than fifty references which provide data on soluble, exchangeable, available, or extractable trace metals in serpentinitic soils and he points out that it is virtually impossible to come up with a consensus because so many different extraction and measuring techniques are used and so many contradictory results have been published.

It is critical to know the proportion of total elemental content that is available to plants. Mineral acids, acetic acids, oxalic acids, ammonium nitrate, ammonium acetate, EDTA, DTPA, ammonium chlorite and water extractions have all been used to determine metal concentrations available to plants, but so far there is no consensus as to what extract best reflects the conditions in the soil. Acetic acid extracts are more widely used and although it is not necessarily the most appropriate extract it is often used to compare other literature sources. Bioassays are probably the most reliable method but since each plant has different tolerance and mechanisms to cope with metals the choice of plant becomes difficult. More details on the topic of plants' response to such soil will be provided in Chapter 5.

What is clearly evident is that the total elemental concentrations of these trace metals reach unusually high values on serpentine soils and Table 14 provides a simple comparison of metal concentrations of selected serpentinitic soils with those not influenced by serpentinites. Nickel is usually the most abundant trace element, followed by manganese, chromium, and cobalt, and the differences in these metals are quite striking. The use of total elemental values to measure toxicity is very questionable but some authors have found significant correlations between total trace metal concentrations in the soils and extractable metal concentrations in plants (Timperley et al. 1970).

A crude index of availability can be obtained from Table 15 which lists the results of studies where both the total elemental and extracted metal concentrations were provided. From this table it is evident that the so-called available portion of nickel, chromium and cobalt is very low in most soils. Nickel is generally more available than chromium and as shown in Table 15 the proportion of the total nickel that can readily be extracted is usually less than 5%. Proctor (1972) also showed that forms of labile nickel might be different from one serpentine soil to the next. In the more organic rich serpentines a significant portion of "available" nickel appears to be associated with organic complexes (Crooke 1956), but some of these complexes are not always available to plants (Halstead et al. 1969). Nickel was found to be available (EDTA and Ammonium acetate extractions) in significant portions in the humus fraction of serpentine soils in Switzerland (Haab 1988). Both the total nickel and chromium content in the humus rich H-horizon were significantly higher than in the fermented LF-horizon and showed enrichment with regard to nickel and chromium levels in the vegetation. Haab (1988) suggested that atmospheric input of serpentine rich dust might be the result of this enrichment. The role of organic matter in nickel toxicity is clearly of importance and needs to be investigated in greater depth.

Table 14

Comparison of trace metal concentrations in serpentine and non-serpentine soils

Location	Source	Co	Ni	Cr	Mn	Cu	Zn	Fe
					ug/g^{-1}			%
Oregon:	(1)	210	2400	900	1900	100	90	11.6
	(1)	130	1100	2700	2900	150	170	13.6
Quebec:	(2)		2000	3000				3.9
	(3)		2000	1000				8.5
Australia	(4)		1900	3400	1300	200		2.4
	(5)	778	3410	634				
Brit. Columbia	(6)	92	1859	433	906			5.6
Newfoundland	(7)	165	3462	1073	1199	26	65	8.5
United Kingdom	(8)	339	6187	10348	2965			14.8
	(8)	141	1750	2221	2196			8.6
Switzerland	(13)	1820	1740	880				5.8
UICC Chrysotile A	(9)	55	1360	1390	450			1.7
UICC Crocidolite	(10)	10	20	8	833			14.6
UICC Amosite	(10)	13	33	35	13680			14.8
UICC Anthopholite	(10)	16	536	217	545			1.3
Non-asbestos soils:								
USA (average)	(11)	10	100	100	600	20	60	2.6
World (average)	(12)	12	34	84	761	26	60	3.2

Source References:
1 = Alexander (1988)
2 = Moore & Zimmerman (1977)
3 = Bordeleau et al. (1977)
4 = Meyer (1980)
5 = Anderson et al. (1973)
6 = Schreier et al. (1987a)
7 = Roberts (1980)
8 = Slingsby & Brown (1977)
9 = Morgan et al. (1973)
10 = Cralley et al. (1968)
11 = Shacklette & Boerngen (1984)
12 = Ure & Berrow (1982)
13 = Sticher et al. (1986)

Nickel in serpentine soils can be mobilized when in contact with organic soils. At a site where asbestos rich sediments were deposited on organic rich agricultural land, Schreier et al. (1987) were able to show nickel mobility in the serpentine-organic contact layer within a few years after deposition. The leaching effects in the contact zone are clearly evident in Figure 12. Mobilization of manganese is also visible but chromium appears to show a less dramatic effect. Similar results of leaching of nickel and manganese were also shown by Sticher et al. (1986) in Swiss serpentine soil profiles.

TABLE 15

Total and "available" trace metals in serpentine soils (Available metals refer to: 1. 0.5N Acetic acid extraction, or 2. Ammonium acetate extraction.)

Element	Total Content	Acid Extr. Acetic Acid ---------- ug/g^{-1} ----------	Ammonium Acetate Extract.	References
Ni	2400	127	6	Shewry & Peterson (1976)
	3330	222	3	" " "
	1540	nd	7	" " "
	3410	390	159	Anderson et al. (1973)
	3220	137	nd	Menezes de Sequeira (1968)
	2430	132	nd	Lyon (1969)
	3462	nd	470	Roberts (1980)
	1700	nd	11	Sasse (1979)
	6187	49	nd	Slingsby & Brown (1977)
	1750	12	nd	" " "
	5224	nd	9	Jaffre (1980)
	4086	nd	46	" "
	nd	136	nd	Lee & Hewitt (1981)
Cr	1600	2.3	0.1	Shewry & Peterson (1976)
	1800	0.9	0.1	" " "
	2010	nd	0.1	" " "
	4930	1.2	nd	Menezes de Sequeira (1968)
	3670	<3.0	nd	Lyon (1969)
	634	<5.0	<7.0	Anderson et al. (1973)
	1073	nd	Tr	Roberts (1980)
	765	nd	1.2	Sasse (1979)
	25896	nd	3.0	Jaffre (1980)
	6297	nd	5.0	" "
	nd	6.0	nd	Lee & Hewitt (1981)
Co	778	32	<5.0	Anderson et al. (1973)
	170	nd	1.2	Sasse (1979)
	nd	23	nd	Lee & Hewitt (1981)

The pH also affects the nickel uptake in plants. By increasing the soil with liming Crooke (1956) and Halstead et al. (1969) showed a significant reduction in "available" nickel in the soil. The presence of calcium ions also reduces nickel toxicity. Finally, the interactions between nickel and magnesium, and nickel and iron might be of importance in pedogenesis since both processes play a key role in the genesis of many deeply weathered lateritic serpentine soils.

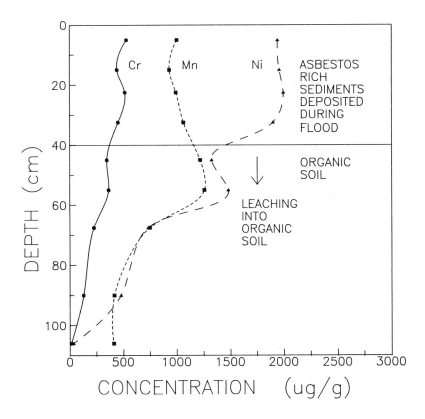

Fig. 12. Nickel, chromium, and manganese mobility and leaching from asbestos rich sediments into organic soils.

Sahu (1981) proposed a nickel fixation model for such soils suggesting that under reducing environments nickel released from parent material, migrate and precipitate as garnierite. This nickel-rich silicate resembles serpentine minerals but nickel has replaced some of the magnesium in the mineral structure (Ogura et al. 1981). Colloidal chemical processes are thought to drive this enrichment process. Under oxidizing conditions nickel is more commonly associated with smectite and limonite. Schellmann (1982) also noted that some serpentines are enriched with nickel and claims that nickel in solution from the limonite zone is fixed in serpentine at the expense of iron and manganese under moderate weathering conditions. In many places these processes result in the formation of high grade nickel ore deposits that are mined commercially in New Caledonia, The Philippines, and Indonesia. Decarreau et al. (1987) showed that up to 50% of the nickel can accumulate in the octahedral layer of smectites, which are generally considered the main weathering product of serpentine soils in humid temperate climates.

High levels of nickel and cobalt have been found to accumulate in a number of plant species on serpentinitic soils. A number of such "hyper-accumulators" have been identified by Brooks (1987) who suggests that some plants evolve to become more tolerant to such metals as nickel and cobalt. The behaviour of cobalt is very similar to that of nickel. Cobalt has been reported to be phytotoxic to plants (Brooks and Malaisse 1985) but no conclusive evidence exists to show that the levels found in serpentinitic soils are high enough to cause plant toxicity. Cobalt appears to be associated primarily with the manganese oxide fraction of the soil and only a small fraction of the total cobalt content could be extracted (Jarvis 1984, McLaren et al. 1986). This suggests that cobalt availability might be limited in many of the asbestos enriched soils.

Brooks (1987) suggests that chromium accumulation in plants is very small and attributes this to the much lower solubility of chromium in soils. Accumulation does not necessarily mean toxicity although many plants with high metal content have a somewhat chloritic appearance and often show early senescence.

Total chromium levels can be very high but because of the low solubility of chromium in nearly neutral conditions there is very little evidence that much chromium is available to plants, nor is there much evidence of chromium uptake in plants. In some podzolic soils on serpentine rock chromium content in the A horizon was found to be lower than in the B horizon suggesting that chromium can be mobile in soil environments where weathering is intense, but tends to bind strongly to clay minerals (Connor et al. 1957). In basic environments chromium is usually present in chromite which is usually resistant to weathering.

High manganese levels have also been reported in some serpentine soils. Manganese functions as an enzyme activator in biological systems and is not very soluble under alkaline conditions.

Availability of manganese increases in poorly drained soils (Swaine and Mitchell 1960) but in spite of the high levels found in some serpentine soils there is no evidence that this element is toxic to plant growth.

Based on the above comparisons of metal content in serpentine soils and different bedrock it can be concluded that most soils influenced by serpentinitic bedrock have unusually high nickel, chromium, cobalt and manganese levels but adverse effects on plants have only been observed with high nickel and possibly cobalt concentrations. The extent of this process is still poorly understood and considerably more research is needed to gain a better understanding of the roles these trace metals play in the soil environment.

4.4 Physical Properties

Much emphasis has been placed on chemical conditions in asbestos rich soils and until the reviews by Krause (1958) and Proctor and Woodell (1975) physical conditions were virtually ignored. Krause (1958) pointed out that serpentine soils often occur in mountainous environments and because of their high porosity and dark colours have a strong influence on micro-climate and moisture regime. Perry et al. (1987) used surface modifications and mulching to enhance water storage and to reduce surface temperatures and evaporation rates in revegetation efforts of asbestos waste sites in the southwestern United States. However, these types of soils show great physical variability and generalizations are not appropriate.

It is surprising that so little attention has been given to the presence of asbestos fibers in soils. Many authors have noted their physical presence, but the role they play in soil genesis, soil nutrition and plant growth has not been investigated to any extent. Since the physical nature of asbestos fibers has been identified by the medical researchers as a factor which causes adverse effects on men and animals it would seem obvious that similar problems might exist for the soil biota and plant roots. This subject merits considerable attention and needs to be investigated in greater depth. All that is known is that some of the asbestos fibers seem to weather under acidic conditions generating considerable amorphous material. The development of colloidal or amorphous silica-type fibers and coatings has been reported from laboratory studies (Sahu 1981, Seshan 1983, and Bellmann et al. 1986) and an indication that such material is also present in serpentine soils has been noted by Istok and Harward (1982). Unfortunately, no other in-depth analysis has been made on fiber durability, fiber alterations and fiber roughness in soils. This topic is relevant in view of the many questions raised in relation to landfill deposition of asbestos waste.

4.5 Asbestos in soils and its effect on soil biology

Little work has been done to document the effects of asbestos rich soils on soil animals. A general reduction in soil animals has been noted in a few studies but this is not surprising considering the general lack of vegetation growth and organic matter supply in such soils. The exception to this is a study by Proctor and Whitten (1971) who observed large pocket gopher populations on a serpentine soil in California. The rocky habitat is well suited to these animals and the presence of a good food source (Brodiaea croms) on these soils might have contributed to creating a more favourable habitat for these animals.

Earthworms are known to tolerate and accumulate trace metals and in an experiment with Lumbricus rebellus Schreier and Timmenga (1986) showed that earthworms not only accumulated nickel and manganese in their body tissue but most of them perished within 21-30 days after introduction into chrysotile rich sediments. pH adjustments only slightly improved survival and the cause of death is postulated to be the abrasive nature of the asbestos fibers. No reproduction or survival was observed in field experiments. This suggests that asbestos fibers have an adverse effect on such animals and both the chemical and physical properties of the fibers might play an important role.

Termites, as documented by Wild (1975) and Brooks (1987), move large quantities of materials from a great depth and in the study on termite mounds on serpentines in Zimbabwe they found increases in nickel, pH, clay, calcium and magnesium in the mounds. The latter three made the mound more fertile for selective grasses and increases in pH are likely to reduce nickel toxicity. According to the same authors termite workers had higher nickel and chromium levels than termite soldiers and this was attributed to different food sources consumed by the different social groups in the termite family.

Information on micro-organism in such soils is also very limited. As reviewed by Proctor and Woodell (1975) there appear to be fewer nitrogen fixing bacteria in such soils, but this might in part be due to the lack of suitable organic substrate in serpentine. White (1967) found significantly fewer nitrogen fixing nodulations in Ceanothus cuneatus on serpentine soils than in adjacent non serpentine soils. He claimed nutrient deficiencies in nitrogen, phosphorus, potassium, and molybdenum, and metal toxicity in nickel and chromium as possible causes for this effect. Fewer numbers of micro-organisms were also observed on serpentine soils in Europe (Ritter-Studnicka 1970). Asbestos rich dust which is deposited on soils from an asbestos mine operation in Quebec not only retarded the soil process of podzol formation (deKimpe et al. 1973) but, as shown by Bordeleau et al. (1977) such additions decreased fungal population and reduced obligate heterotrophic bacteria. At the same time facultative heterotrophic and autotrophic bacteria were increased. Increase in pH and the addition of new readily oxidized materials are considered the reasons for the increase in the facultative bacteria. It is unclear what causes such reduction of microbial activity. Trace metals such as nickel have been found to inhibit the growth of eubacteria, actinomycetes, cyanobacteria, yeast, filamentous fungi, protozoa and algae (Babitch and Stotzky 1983). Given the presence of other trace metals, the complete imbalance of nutrients and the relatively adverse pH of most asbestos rich soils it is unclear at this time whether nickel alone or a combination of factors is responsible for poor bacterial development in such environments.

From these few studies it is evident that asbestos rich soils have adverse effects on soil biology but neither the extent nor the cause of the problem is well documented and as suggested by Proctor and Woodell (1975) and Brooks (1987) this fascinating topic merits more in depth investigation.

4.6 Asbestos rich soils and animal health

Ingested soil plays a significant role in grazing animal nutrition. As reviewed by Thornton (1981) up to 15% of dry matter intake in sheep and 10% in cattle can be soil and these figures can be even higher in the winter. He also suggested that there is often a good relationship between metal levels in the soil and those found in the blood of the grazing animals. Lead, copper and zinc levels in the blood of cattle were shown to be elevated in cattle grazing on pastures in mineralized districts in Britain. A similar trace metal uptake is expected in the blood of animals grazing on serpentine rich soils. It could be argued that such exposure is unlikely since these soils are unsuited for productive pastures, but exposure and uptake are nevertheless possible as documented by Schreier et al. (1986) in the Sumas river basin in British Columbia. During a storm several pastures were inundated with asbestos rich sediment, and subsequent grazing by cattle in the vicinity of chrysotile rich depositions resulted in sufficient inhalation and ingestion that blood samples collected after periods of grazing showed elevated levels of nickel and manganese. These metal levels were only elevated during summer grazing and returned to normal levels in the winter when the animals were fed with imported feed in the barn (Figure 13). Long term effects on animal health could not be determined in this study because the exposure had been too recent and the population too small and genetically too diverse. Nevertheless the metal uptake by the animals could be used as a tracer for the asbestos fiber pathways.

Since lung biopsy is the only sure way of determining long term fiber exposure sheep lungs have been used to provide evidence of fiber exposure in some of the Turkish studies. This area of research is certainly gaining interest if the link between asbestos rich soils and deteriorated human health can be established. A review of such conditions is provided in the section below.

4.7 Asbestos rich soils and human health

It has been shown that asbestos rich soils exhibit generally unfavourable conditions for plant growth. The lack of vegetation cover combined with the small size of the fibers makes serpentinitic soils more prone to wind erosion. People living in the vicinity of such soils are therefore expected to be exposed to significant airborne asbestos fibers. Because of their close association with working the soils farmers are potentially at risk. Fortunately, because of the generally unfavourable growth conditions associated with serpentine soils, few such soils are cultivated. Evidence of possible asbestos fiber contamination in agriculture as a result of working such soils is still fragmentary but there are signs that some problems exist in a number of areas in eastern and southeastern Europe (Wagner 1986).

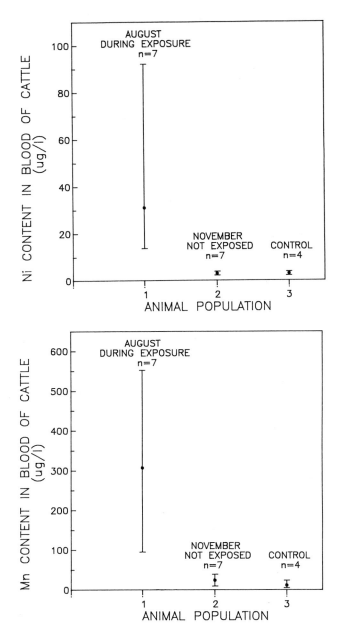

Fig. 13. Nickel and manganese levels in blood of cattle grazing in fields affected with asbestos rich sediments. (1 = Animals grazing in affected area, 2 = Animals removed from pasture and feed imported unaffected feed, 3 = Control animals with no exposure).

The first indication of rural medical problems came from Finland
where Kiviluoto (1960 & 1965) observed pleural plaque in the rural
population living in an area with high concentrations of the
amphibole anthophyllite in the bedrock. Burlikov and Michailova
(1970) found a link between cultivating soils rich in amphibole
asbestos (anthophyllite and tremolite) and pleural plaque in the
farming population of Bulgaria. Recent studies have indicated that
the problem is complex because not only asbestos fibers but other
naturally occurring fibers such as erionite (metamorphically
altered volcanic material) might also be responsible for producing
mesothelioma in the rural environment (Wagner 1986, Wagner and
Pooley 1986). Evidence of possible cause and effect relationships
have come from lung biopsy studies where erionite, rutile,
actinolite, tremolite, crocidolite, amosite and chrysotile fibers
have been found in lung tissue of sheep and rural settings. The
same authors claim that fine tremolite rather than chrysotile
fibers found in many lungs of workers involved with chrysotile
mining and manufacturing might play a key role in the development
of asbestos related diseases. Both types of fibers seem to be
present in most chrysotile asbestos and a separation of the two is
virtually impossible. The durability and acid resistance of these
two fibers are however very different.

Indication of adverse health effects in fiber rich non-
occupational environments have recently been reported from a number
of different areas. Baris (1975 and 1980) found a high incidence
of mesothelioma in a Turkish village where the soils and bedrock
contain large quantities of very thin erionite fibers. Wagner and
Pooley (1986) have shown that unusually high incidence of
mesothelioma was observed in test animals exposed to these fibers.
However, it should be noted that Rohl et al. (1982) showed that
chrysotile and tremolite fibers were also present in the Turkish
dust samples.

Similar links between fibers present in soils and rocks and
human health problems in the rural villages have been reported by
McConnochie et al. (1987) in Cyprus, by Constantopoulos et al.
(1985 and 1987a&b), Langer et al. (1987) in Greece, and by
Yazicioglu et al. (1980) in Southwest Turkey. In all these cases
chrysotile and tremolite fibers have been identified in lung
biopsy. Another interesting study on the island of Corsica has
shown a significantly higher incidence of pleural plaque in people

living in areas near asbestos deposits (Boutin et al. 1986). Finally, Churg and dePaoli (1988) also showed incidence of pleural plaque in farmers in the asbestos mining areas of Quebec. Again chrysotile and tremolite fibers were implicated with the disease in both studies. A summary of the relevant literature is provided in Table 16.

TABLE 16
Areas with potential rural health problems associated with the presence of asbestos fibers in local soils and bedrock

Location	Activity & Environment	Fiber Type	References
Turkey	Whitewashing and airborne dust	Erionite (Zolite) Chrysotile Tremolite	Baris (1975 & 1980) Baris et al. (1978 & 1979) Artvinli & Baris (1979) Baris et al. (1987) Yazicioglu (1976) Yazicioglu et al.(1980) Rohl et al. (1982) Langer et al. (1987)
Greece	Whitewashing and airborne	Tremolite	Constantopoulos et al. (1985 & 1987 a & b) Bazas et al. (1981)
Cuprus	Environmental	Chrysotile Tremolite Crocidolite	McConnochie et al. (1986)
Corsica	Environmental	Chrysotile Tremolite	Steinbauer et al. (1987) Boutin et al. (1986)
Quebec	Environmental near mining	Chrysotile Tremolite	Churg and de Paoli (1988)
Finland	Environmental near mining	Anthophyllite	Kiviluoto (1960 and 1965)
Bulgaria	Agriculture	Tremolite Anthophyllite	Burlikov & Michailova (1970 & 1972)
New Caledonia	Mining & Environmental	Chrysotile Tremolite	Leclerc et al. (1987)

From these studies it becomes clear that asbestos fibers in the soil can be implicated with human health in a number of cases. The fiber size in such soils seems to be of particular importance. How much individual fiber types contribute to the problem of pleural plaque and mesothelioma development is still uncertain and it is hoped that future projects will enlighten us in determining exposure hazards. What is clearly needed from the pedology community is a better quantification of the physical size, properties and durability of fibers in the soil medium.

4.8 Improving soils to enhance revegetation

Since most asbestos rich soils are notoriously infertile and have many nutritional disorders, relatively little effort has been made to alter the conditions to produce an environment more conducive to plant growth. However, concern has been expressed in relation to mine reclamation, quarrying, the presence of serpentine soils in the vicinity of urban expansion, land reclamation of sites containing asbestos fiber waste, and deposition of sediments rich in asbestos fibers along stream banks.

There are two major concerns at such sites: 1) emission of asbestos fibers by wind and water during the extraction, processing and deposition of asbestos materials, and 2) subsequent emission by weathering and erosion at such sites. The extent of asbestos fiber distribution in the airborne environment will be reviewed in Chapter 6 and it is believed that fiber input into the atmosphere can be significantly reduced if the source areas can be covered with adequate vegetation growth. Some of the rock dumps and mine tailings in the Quebec area go back some 60 years and if left on its own, even after that length of time vegetation establishment is effectively non-existent. This suggests that soil and fertilizer amendments are needed to speed up the revegetation process.

Minimal effort has gone into the reclamation of asbestos rich mine tailings. To turn such sites into productive agricultural and forestry land is a very difficult task and the main issue in such reclamation is simply to establish and maintain a protective vegetation cover. This conservation measure helps reduce asbestos fiber erosion and distribution by wind. In all such cases the addition of topsoil has proven to be necessary since asbestos wastes have a wide range of nutrient deficiencies and components with potential toxicity.

Results from asbestos mine tailing and waste reclamation efforts in Australia (Meyer 1980), Canada (Moore and Zimmermann 1977, 1979) and the United States (Perry et al. 1987) indicate that the addition of top soil cover is critical for germination and to overcome the adverse physical and chemical conditions in the long run. The low water holding capacity of such mine waste material is a major problem for seed germination and plant growth during dry periods. Mulching proved necessary to establish vegetation on asbestos waste in the semi-arid southwestern United States (Perry et al. 1987). Moore and Zimmermann (1977, 1979) had best success with fertilizer additions in combination with farmyard manure, sawdust, and sewage sludge and were able to establish grasses and forbs in their reclamation work in Quebec. The deficiencies in some of the major nutrients are partially overcome by the addition of nitrogen, phosphorus and potassium fertilizers in combination with a calcium source. However, the form in which calcium is applied appears to be critical. Most asbestos waste material has a very alkaline abrasion pH and the addition of lime might increase the pH even further. Under these conditions the metal toxicity is likely reduced but at the same time phosphorous availability might be impaired. The application of gypsum as a neutral source of calcium proved to be a most effective way to overcome calcium deficiencies and poor Ca/Mg ratios (Meyer 1980). Moore and Zimmermann (1977) also used gypsum in combination with nitrogen, phosphorus and potassium fertilizer and organic matter. In spite of using very high application rates, nitrogen and calcium deficiencies were observed in the second year. From their work it is apparent that it is not only difficult and expensive to establish vegetation on such sites but it might be equally difficult to maintain the cover.

With increasing additions of high levels of fertilizer salinity problems develop and as the pH drops metal toxicity might become apparent over time. The choice of vegetation species is important and will be discussed in greater detail in Chapter 5. The use of established serpentine flora is an obvious choice but seeds of such vegetation mixes are very difficult to obtain in large quantities. Given that there are more than 5.5 km^2 of asbestos tailings in Quebec alone it seems surprising how little effort has gone into research on mine reclamation. The unusual conditions of this material make it obvious that conventional mine reclamation methods are inadequate and a more focused approach is needed. One of the most likely reasons that so little effort is put into asbestos mine reclamation work is that it is expensive because top soil and organic rich waste might not readily be available and transport costs for top soil, organic materials and fertilizers can be substantial. In the Canadian context Moore and Zimmermann (1977) and Jolicoeur et al. (1984) have given figures of $ 0.30/m^2 to $ 0.50/m^2 for reclamation and revegetation costs of asbestos mine tailings in Quebec. Hydroseeding is generally recommended for steeply sloping areas.

In some studies vegetation response was very good but in the majority of cases response was small or negligible. With high soil input initial revegetation success is good but long term deficiencies appeared several years after the first treatment, and it appears that continuous additions of soil amendments are necessary.

4.9 <u>Modifying asbestos waste in the soil</u>

Most asbestos waste is dumped and buried in the soil and the fate of this waste is of concern since it is toxic and unlikely to disintegrate very rapidly under normal soil forming processes. Probably the major concern is the resuspension of such waste into the atmosphere. Once the waste is buried and the site is covered with vegetation this process can be minimized. Any disturbance is to be avoided and agricultural production is not recommended since plowing could bring some of the waste material to the surface. Agricultural production is likely to be limited due to the adverse nutrient problems associated with such waste. The problem of fibers seeping into the water supply is also of some concern but the soil is likely to act as a filter. A soil cover of more than 1 m was recommended by Cook and Smith (1979) in order to avoid surface disturbance by vehicles and asbestos exposure by rain and surface erosion.

Several suggestions have been made that chrysotile rich waste should be treated with waste acids prior to disposal or at the disposal site. This was prompted by evidence that chrysotile fibers appear to be significantly less hazardous when leached with acids (Morgan et al. 1977, Monchaux et al. 1981, Jaurand et al. 1984, Harvey et al. 1984, Jaurand et al. 1988). Tests to this end have been carried out by Baldwin and Heasman (1986) in Great Britain and the results showed that chrysotile could be leached efficiently with strong mineral acids and to some extent with carboxyl acids. This is however not a simple procedure and in the case of organic leachate acids the process is slow. What was not pointed out in their experiments is that in addition to magnesium, trace metals present in high concentrations in the asbestos fibers are also released. Evidence of this is given by Schreier et al. (1987a & b) who showed trace metal release from asbestos rich stream sediments into the underlying organic rich soil horizons. In the same environment trace metals are released into the stream water due to weathering and acidification (Schreier 1987). In addition, tremolite fibers, which appear to be present in most chrysotile asbestos material and which, according to Wagner (1986), are likely the key fibers responsible for human health problems, are acid resistant and thus remain unaffected by acid treatment.

If the leaching process is unaided by man it will take place in the soils but at a very slow rate. The topic relating to waste disposal will be discussed in more detail in Chapter 6.

4.10 Summary of asbestos fibers in the soil environment

Soils developed on serpentinitic bedrock are known to contain significant portions of chrysotile asbestos. These areas are small in size but have a worldwide distribution. They are characterized as having very unbalanced nutrient conditions and excess trace metal content. Pedologists have almost entirely focused on improving the calcium, nitrogen, phosphorous, potassium, and molybdenum deficiencies and reducing the potential toxic effects by excess nickel, chromium, cobalt, manganese and magnesium levels in these soils. Yet, physical properties which are so prevalent in medical research has so far been ignored.

It appears that some of the fibers leach and weather into clay minerals, preferentially into the smectite category. Soil biology and microbial activity also appear to be adversely affected by these soils although only a few studies have been carried out so far. Fragmentary information exists to suggest that some of these asbestos rich soils have an indirect effect on animal and human health, and the amphibole fibers more so than chrysotile fibers appear to be implicated with the health problem. In this context it is evident that the physical properties, the durability and weathering process of asbestos fibers in the soils need to be investigated more thoroughly.

Adverse chemical and physical conditions make reclamation work on asbestos rich mine tailings and disposal sites difficult. The establishment of a plant cover is not simple. Additions of calcium, adjustments of pH, and applications of NPK fertilizers were partially successful, but tolerant plant species are needed in order to establish a permanent vegetation cover on such sites. Acid leaching of fibers and subsequent land fill application with deep soil cover have been proposed as a suitable disposal method but questions of surface disturbance, wind erosion and trace metal release at such sites remain of concern.

Soils developed on undisturbed asbestos rich bedrock can serve as a key for reclamation work, but the diversity of the conditions, the durability and small size of the fibers and the many other characteristic factors of serpentine soils are likely to mystify and challenge pedologists in times to come.

CHAPTER 5

ASBESTOS AND PLANT GROWTH

5.1 Introduction

As noted by Kruckeberg (1969) vegetation on ultramafic bedrock represents one of the clearest examples of edaphically induced vegetational discontinuity. Climate is generally considered the most important factor influencing plant growth but in the case of serpentinitic rocks and soils the edaphic control of plant distribution indicates that the geological factor might be of equal importance to the climate. Although not widely documented most serpentine rocks contain asbestos fibers and the effects of serpentine rocks on plant distribution and growth has important and direct relevance to all materials and soils containing asbestos.

It is generally acknowledged that plants have a hard life on asbestos bearing serpentinitic soils and botanists have long studied the reasons for their unique and unusual plant distribution. As noted by Proctor and Woodell (1975) serpentine habitats are often regarded as museums of plant evolution because they harbour rare plants, endemics and species of disjunct distribution. Vegetation on serpentine soils is usually in stark contrast with the more luxuriant plant distribution in the surrounding non-serpentinitic areas. Serpentine vegetation is characterised by xerophytism, depauperization, dwarfism and unique physiognomy and taxonomy (Kruckeberg 1969).

Much has been written on serpentine vegetation and the following references are excellent reviews of the subject: Brooks (1987), Kruckeberg (1984), Proctor and Woodell (1975) and Krause (1958), and Whittaker (1954).

The first part of this chapter is devoted to a description of plant distribution on asbestos rich soils and the origin and evolution of endemic plants. The second part discusses responses and adaptations of plants to such adverse conditions as moisture stress, excess magnesium and trace metals, and nutrient deficiencies. Finally, a discussion is provided on plant selection used in revegetation efforts associated with the reclamation of asbestos mine waste and disposal sites.

5.2 Plant distribution on asbestos rich serpentinitic sites

The plants most frequently found in serpentinitic environments have been characterised by Brooks (1987) as belonging to insular (neoendemism) and depleted taxa (paleoendemism), although it should be noted that such a distinction is not easy to make. Residual serpentinitic soils, developed in the tropics on undisturbed bedrock, appear to support a richer plant community than serpentinitic soils in North America and Northern Europe where glaciation has disturbed the residual soil environment and retarded plant evolution. The African and Asia-Pacific serpentine flora had more than 80 million years to develop and adjust to the adverse physical and chemical conditions of the serpentinitic substrata. This is in stark contrast to the 10,000 years of evolution in landscapes subjected to recent glaciation. As a result the tropical serpentine flora contains a significantly larger number of species than those found in glaciated environments. Kruckeberg (1984) found more than two hundred and fifteen species, sub-species and varieties of plants in California alone that are restricted at least in part to serpentine. In contrast, Kruckeberg (1979) suggested that there are very few species which are completely restricted to serpentine in British Columbia, where glaciation was much more extensive. Considerable differences exist between species which occur both on and off serpentine materials. In tests Kruckeberg (1954, 1967) found that species evolved on serpentine were considerably more tolerant to low levels of calcium and other stress than the same species that evolved on non-serpentine soils. Of those species occurring in both serpentine and non-serpentine soils it was found that the non-serpentine species were more vulnerable to competition by weeds and other plants. Finally the same author found that certain non-tolerant species occurring on adjacent sites grew better on serpentine soils than serpentine tolerant species from base rich sites a long distance away. This suggests a considerable genotypical specialization.

The tropical flora also has more distinct physical and chemical characteristics and many species have lost their closest relatives on non-serpentinitic environments due to depletion. In contrast, in glacial environments many species can be found both on and off serpentine sites but the plants on serpentine sites often show stress symptoms such as reduced growth and lower frequency. Finally, more distinct taxa are found in separate serpentinitic environments in the tropics than in the areas influenced by glaciation.

Botanists have likely paid more attention to serpentine sites with very contrasting vegetation than to sites with few distinguishing characteristics. This then provides a somewhat misleading picture because there are many cases where the differences between serpentine and adjacent non-serpentine vegetation are not very striking (Proctor and Woodell 1975). Whittaker (1954) found some of the serpentine vegetation to be stable and self-maintaining. This is somewhat contrary to the view that given enough time for mature soil development vegetation will converge to a climatic climax condition. Both Whittaker (1954) and Proctor and Woodell (1975) consider some of the plant communities on serpentine to be at their unique climax state.

Coniferous trees appear to be able to tolerate serpentinitic conditions far better than most broadleaf trees, and coniferous trees, sclerophyllous shrubs, grasses and forbs are the most common plant types on many serpentine soils. Many good indicator species exist but they are usually restricted to floristic regions of the world.

It is somewhat surprising that serpentine sites often harbour a greater number of species than adjacent non-serpentine soils (Proctor and Woodell 1971). The development of mature forests usually restricts plant diversity and since vegetation cover is open and tree development impaired diversity seems to be somewhat enhanced on many serpentine sites. However, in extreme conditions both plant diversity and plant cover are so restricted that the sites are barren, a condition very common in serpentine rich environments.

One of the most characteristic features of vegetation across serpentine boundaries is a shift in vegetation type and morphology. Kruckeberg (1984) noted that oak woodlands are replaced by chaparral, Douglas fir and hardwoods give way to pine-cypress, and chaparral is replaced by grassland on serpentine sites in California. Xerophytic shrub replaces tropical rain forests in New Caledonian serpentine, grassland replaces some tropical forest in Zimbabwe and dense beech forests change into stunted shrubland or tussock grass in other serpentine areas (Proctor and Woodell 1975). Kruckeberg (1984) further documented that Jeffrey pine, Incense cedar and various firs are usually present on serpentines in

California, while Redwoods, Sugar and Yellow pine and True fir were absent.

5.3 Serpentine Induced Stress and Plant Response

As highlighted in Chapter 4 asbestos bearing soil makes an unusual plant habitat. There is great internal variability within such sites and large contrasts exist from one serpetinitic site to another. Moisture stress, excess magnesium, low Ca:Mg ratios, excessive nickel, chromium, and manganese levels, and deficiencies in molybdenum, calcium, phosphorus, and nitrogen have all been cited as key factors responsible for poor plant growth. Since many interact it has been impossible to single out any one of them as the key to limiting vegetation growth.

Plants that can survive on such materials cope with the unusual conditions in many other ways than simple species selection and species adaptation. Probably the most obvious expression of stress is in the plant morphology where three effects are most common: 1) Xeromorphic foliage with changes in coloration; 2) Reduction in size with shrubbiness, stunting, and plagiotropism; and 3: Development of extensive root systems.

The plants can respond in many ways to the physical and chemical stress factors induced by asbestos rich materials. Excess cations can be excluded or restricted, deficient cations might be taken up in a preferential manner, excess metals might be taken up and stored in a different compartment of plants, or uptake might be restricted altogether. As noted by Lyon (1971), Kruckeberg (1984) and Brooks (1987), there is no universal process by which plants survive on serpentine soils. Instead plants adjust to the conditions in their own individual way and a large variety of complex processes are involved.

5.3.1 Physical stress

Many serpentine soils are shallow and stony which results in a poor water-holding capacity. Although most asbestos material is light in colour the host serpentine rock is dark thereby absorbing more energy leading to greater diurnal fluctuation, which in part can affect the aridity on such sites. The moisture stress might be one of the causes of serpentine plants developing a more extensive root network than other plants. It appears that xeromorphic vegetation survives better on dry serpentine sites but there is enough variability for Proctor and Woodell (1975) to suggest that such physical problems alone are not sufficient to account for the unusual plant response. In contrast, some serpentine soils are poorly drained and highly susceptible to landsliding, which is a very frequent occurrence on serpentines.

In view of the overwhelming medical evidence that the physical properties of asbestos fibers are hazardous to health, it is very surprising that no investigation has been carried out by botanists and plant ecologists to document whether individual fibers have an adverse effect on roots and whether such fibers, which are known to penetrate human cells, actually penetrate plant cells. It can be postulated that entry of fibers into the root system could adversely affect plant growth. However no evidence of any research on this topic could be found and it is suggested that this might be an enlightening research topic. Another area of interest is the soil root interface. Do these tiny asbestos fibers, which are known to have very sharp ends, alter root development? Are root injuries more frequent in asbestos soils than in the adjacent area?
There is no direct evidence that this is happening but the topic deserves closer attention since there appears to be a significant reduction of micro-biological processes in serpentine soils (Proctor and Woodell 1975).

5.3.2 Chemical Stress

Stress can be exerted by either excess concentrations or large deficiencies. Much has been written about the inverse relationships between calcium and magnesium levels in asbestos rich material. In most normal soils calcium generally is far more abundant than magnesium. In serpentines calcium levels are often extraordinarily low while magnesium levels above 20% are not uncommon. Numerous authors have shown that some plants restrict magnesium uptake beyond the roots by storing magnesium in the non-living cells of the roots. Other plants exclude magnesium uptake altogether, and in some the magnesium requirements were shown to be much higher than in cultivated or adjacent plants. Brooks and Xing-hua Yang (1984) found inverse relationships between magnesium and aluminum, boron, cobalt, manganese, phosphorus, and sodium and from these observations they concluded that magnesium uptake in plants might have an antagonistic effect and reduce the uptake of the other important nutrients. Magnesium also plays a significant role in trace metal uptake and will be discussed below.

It has often been suggested that calcium deficiency plays a key role in serpentine infertility but it is now clear that many other factors are also very important. Some plants accumulate calcium in a preferential manner (Walker 1954), others have adapted to low calcium values by having more efficient uptake mechanisms (Main 1974). The majority of plants introduced to serpentine soils or asbestos waste respond favourably to calcium additions but the form in which calcium is supplied appears to be critical. Some native serpentine plants show excellent growth at high magnesium concentration but yields are reduced as the Ca:Mg ratio reaches 2:1. In contrast, the same plants from non-serpentine sites increase in growth up to a Ca:Mg ratio of 1:8. Main (1974) concluded that root morphology, uptake mechanisms, translocation of nutrients, and interactions between cations all play important roles in serpentine plant growth and calcium and magnesium cycling. Finally, it should be mentioned that normal Ca/Mg ratios in native plants on serpentine soils are not uncommon (Wallace et al. 1982).

Nitrogen, phosphorus, and molybdenum deficiencies are very important and plant response is usually improved if all three are applied in combination with calcium.

Probably the most interesting components involved in stress are excess metals such as iron, nickel, chromium, and cobalt. In poorly drained acidic sites iron might become toxic, because ferrous ions are more soluble and might reach toxic levels. However, the pH levels in serpentine are most often in the neutral to alkaline range, hence solubility is reduced. Nevertheless, some serpentine plants have shown very high levels of iron (Proctor and Woodell 1975).

Traditionally much attention has been paid to plant toxicity induced by excess trace metals. Metal accumulation by plants has recently become a topic of considerable interest to geologists (Cole 1973, Horler et al. 1981). The term " hyper-accumulator ", proposed by Brooks (1987), has been used to describe many serpentine plants that are enriched in nickel to a level far beyond that found in the soils.

Many authors have reported that selective species are capable of accumulating nickel (e.g. Minguzzi and Vergnano 1948, Severne and Brooks 1972, Wild 1970), but Brooks and his co-workers were able to produce the most comprehensive list of hyper-accumulating

plant species in the world by analyzing over 2000 herbarium plants
(Brooks et al. 1977, Wither and Brooks 1977). A complete review of
the subject of nickel hyper-accumulator plants is provided by
Brooks (1987, Chapter 8) who defined hyper-accumulators as plants
that reach levels above 1000 ug/g of nickel on a dry weight basis.
This is in contrast to most other plants on serpentine soils which
reach levels of up to 150 ug/g (Sasse 1979). As shown by Brooks
(1987) the class limits for hyper-accumulators is not arbitrary but
clearly shows a polymodal frequency distribution of nickel
concentrations for one major plant group (Figure 14).

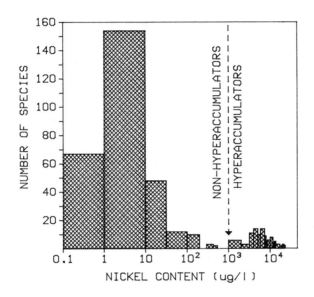

Fig. 14. Frequency distribution of plants capable of accumulating
nickel. (Source: Brooks 1987, Dioscorides Press Ltd, Portland,
Oregon, with permission).

Brooks (1987) suggests that all the hyper-accumulator plants
occur on ultrabasic rocks and although it is not stated
specifically it is assumed that many if not most of the soils on
these sites contain asbestos fibers. The relationships between
hyper-accumulators and ultramafic rocks is so good that Wither and
Brooks (1977) and Brooks et al. (1977) were able to pinpoint
previously unknown ultrabasic rocks in Southeast Asia from analysis
of herbarium plants.

Almost all hyper-accumulator species occur in warm temperate and tropical environments such as New Caledonia, Southeast Asia, Zimbabwe, Cuba, Australia, and Southern Europe-Asia Minor. No such plants have yet been identified in glaciated environments, and only two species (Streptanthus polygaloide and Thlaspi montanum) appear to be present in North America (Reeves et al. 1983, and Reeves & MacFarlane 1981). The genus Alyssum and Thlaspi are the most important hyper-accumulators of nickel in Southern Europe and Asia Minor (Reeves and Baker 1984, Brooks 1987). Almost one half of all reported nickel hyper-accumulators originate in New Caledonia where the greatest endemism is found in the serpentine flora. Zimbabwe is another area with a high occurrence of nickel hyper-accumulators.

Brooks (1987) noted that of the 144 identified hyper-accumulators of nickel, 95 belonged to the Brassicaceae and Flacourtiaceae families, and this suggests very restricted evolutionary adaptation to the unusual soil conditions.

As shown by Shewry and Peterson (1976) trace element concentrations in plants growing on metal enriched sites vary greatly from species to species and due to the lack of a reliable extraction technique total elemental content is still the best measure of "availability". In most serpentine plants higher trace metal levels were found in roots than in the shoots or leaves (Sasse 1979b) and the main factors influencing plant uptake are soil pH, presence and concentration of other ionic cations (Hughes et al. 1980). Old roots tend to accumulate inactive metals and the same author reported that the cortex of the root of European serpentine plants accumulated six times more nickel, two times more cobalt, and 13 times more chromium than the wood. This is in stark contrast to hyper-accumulators where the highest levels are found in the leaves, and the lowest in the roots. These plants appear to have developed a mechanism by which they tolerate, translocate, and store excessive amounts of nickel in the leaves, and the nickel is then released from the plants through leaf fall. In many plants the litter and bark are generally the main plant components with enhanced metal concentrations, and the trace metal concentrations of the whole plant usually increase with maturity, and seasonal variations have also been observed (Hughes 1981).

The metal uptake mechanism in plant metabolism is still poorly understood (Peterson 1983). As far as nickel metabolism is concerned, Lee et al. (1977, 1978) have shown good relationships between citric acid and nickel in New Caledonian plants. Malic and malonic acids were also found to play an important role in the nickel metabolism of several plants. They suggest that some of these plants have the capacity to produce organic acids that form complexes with nickel outside the root membrane and then transport such complexes to the leaves where a large portion of the nickel is tightly bound to the cell walls in the vacuole. This prevents free aquonickel(II) ions, which are phytotoxic, from entering the plant.

The chromium level in plants is usually well below that in the soils and phytotoxic chromium was not accumulated in serpentine plants. In contrast to man, chromium is not an essential element to plants. Lyon et al. (1969) reported that most chromium entering the plants is retained in the roots in the form of a chromium-oxalato-complex and this reduces the uptake of phytotoxic chromium. No hyper-accumulators of chromium have yet been found but a few chromium accumulating plants have been identified by Lyon et al. (1968) who suggested its use in geobotanical prospecting. Uptake of chromium through the roots is very slow and a very small proportion accumulates in shoots (Shewry and Peterson 1974). Chromium apparently enters the vascular tissue with difficulty but once there it can be transported rapidly (Peterson and Girling 1983). The form of chromium in plants is of considerable importance in animal and human nutrition both as a nutrient and as a toxin.

Brooks (1987) reported several cobalt hyper-accumulators. Yang et al. (1985) reported some interesting inter-elemental relationships in serpentine plants. Cobalt was inversely correlated to boron and sodium, and nickel to boron, sodium, and manganese. This could suggest that cobalt and nickel might inhibit the uptake of some of the nutrients mentioned. As discussed by Peterson and Girling (1983) legumes and cereal forages are generally high in cobalt, while rye and orchard grasses are low. Cobalt appears to accumulate in marginal areas of clover and primrose leaves and in the extreme tips of some grasses. Since cobalt is relatively non-toxic to animals and plants the concern in relation to asbestos is low.

Although many of the above mentioned plant processes are still poorly understood, the result shows how nature has adjusted to adverse conditions in an evolutionary manner. Additional research is needed, however, to determine what role, if any, asbestos fibers play in preventing normal plant growth. Do these fibers interfere with plant metabolism, do they penetrate into the root and plant system, and are some plants able to selectively exclude fibers? Is plant root expansion inhibited by the presence of sharp and abrasive needles ? I have found no answers to these questions but such processes are of great interest to people involved in reclamation work of asbestos mine and industrial waste since revegetation of such sites has proven to be such a challenging undertaking.

5.4 Plant selection for revegetation of asbestos waste materials

There are many asbestos rich waste sites in many countries around the world, and revegetation of such sites has been difficult. With population growth and few land use restrictions many of the asbestos rich mine and waste sites as well as natural areas containing asbestos bearing soils will become a greater potential health hazard. Establishing a vegetation cover on such sites will at least reduce the spread of asbestos fibers by wind action to some extent. Since natural colonization is very slow reclamation of the sites is the only alternative. Attempts to revegetate mine waste sites have been reported by Moore and Zimmermann (1977, 1979), Meyer (1980), and Perry et al. (1987).

The dilemma is whether to use commercially available plants for revegetation or serpentine tolerant species. The former are much more readily available but require major soil amendments and fertilizers to establish them and to maintain their growth on asbestos rich sites. Serpentine tolerant plants would require less input and maintenance but seeds and seedlings are very difficult to obtain. In any case none of these sites are suited for productive agriculture and forestry since the input would be too great for competitive market production.

The best solution, that is to apply a thick soil cover, is very costly. Instead, mulching, manure, sewage and other organic rich waste applications have shown some promise in creating a more favourable medium for plant growth (Moore and Zimmermann 1979, Jolicoeur et al. 1984). Given the importance of climate as a factor in plant growth, species which are tolerant to the prevailing climatic conditions should be selected. That in itself is not sufficient since experiments have shown that although climatically tolerant commercial plants could be established at many sites, long term sustainability proved to be very difficult and expensive. Moore and Zimmermann (1977, 1979) used perennial ryegrass, Canada and Kentucky bluegrass, and sweet clover as major plants for their revegetation efforts on asbestos mine waste in Quebec. The addition of Russian wildrye grass, smooth brome and white sweet clover was also somewhat successful. These plants are native to the Canadian Prairie and have adapted to such adverse conditions as high pH, high base cations, and drought, all of which are present in asbestos tailing sites. The same authors concluded that acceptable vegetation cover over large areas could only be obtained by inorganic fertilization combined with organic amendments. A wider range of species and continuous applications of fertilizer are needed to arrive at a permanent vegetation cover and we are a long way from establishing self maintaining and sustainable vegetation at low input rates on asbestos tailings. Meyer (1980) used barley (Hordeum vulgarie) in his pot experiments with asbestos mine waste in Australia, and he had considerable difficulty establishing good root growth in spite of many different fertilizer treatments.

Attempts to improve rangelands in serpentine environments in California have been carried out to replace low palatable species with forage grasses (Kruckeberg 1984). The results have been somewhat discouraging since selective grazing by wildlife removed many introduced grasses, and phosphorus and nitrogen deficiencies remained in spite of fertilizer applications. Kruckeberg (1984) recommends that some of these lands be placed under protection to preserve rare serpentine species since these soils cannot easily be converted into productive agriculture.

Crops such as oats, alfalfa, beans and cabbage have poor tolerance to asbestos rich serpentine soils (Hunter and Vergnano 1952, Mizuno 1979, Brooks 1987). Anderson et al. (1973) described symptoms of metal toxicity in oats and Halstead (1968) suggests that organic amendments served as an agent for complex nickel in the same plant. Proctor and McGowan (1976), found that high magnesium levels reduced nickel toxicity symptoms in oats. This latter process seems to be species specific because no amelioration of nickel toxicity by high magnesium levels was observed in experiments with Festuca rubra by Johnston and Proctor (1981). Similarly, differences in species response were also observed with different calcium applications to such sites (Proctor and Cottam

(1982).

Timothy, ryegrass, orchard grass, potato, maize and rice seem to be more tolerant in some of these soils (Brooks 1987, Mitzuno 1979) but in most crop experiments large soil amendments were necessary to establish adequate growth and commercial agriculture is not recommended on such sites.

The establishment of trees on mine tailings has proven to be very difficult. Seedling survival in the droughty and chemically adverse conditions prevalent in the tailings is the main problem initially. By additions of organic matter and soils trees can be established but once the roots expand into the asbestos substrate the conditions are very detrimental to the trees. Moore and Zimmermann (1979) claim that only willow trees survive in their experiments and the growth rate of such trees is significantly impaired.

Ernst (1976) distinguished between external tolerance mechanisms to prevent metals from entering plant tissue and internal mechanisms where plants take up metals but translocate them to avoid sensitive pathways and metabolic sites. It would be appropriate to select plants that exclude metal uptake for reclamation but as shown by Taylor and Crowder (1983) such experiments are difficult since the behaviour of some of the plants grown in cultures and under field conditions show contradictory results. Typha latifolia grown in solution cultures were unable to restrict entry of nickel and copper, yet field tests indicate that these plants effectively restrict metal intake.

The use of seeds and plants native to serpentine sites is desirable because planting species that are tolerant to these unusual soils assures long term survival. However, the solution is not a simple one. Unlike the reclamation of other metalliferous sites (Smith and Bradshaw 1979) there are no commercially available cultivars tolerant to asbestos waste. Secondly, native serpentine plants do not grow very vigorously and do not respond in an expected fashion to amendments. Thirdly, serpentine plants that grow on and off serpentine sites have been shown to have different genetics and respond differently to soil amendments and fertilizers (Brooks 1987).

The most expedient method for vegetation establishment is to cover the waste sites with fill, non-serpentinitic soils, and organic matter to a depth greater than the rooting depth of the plants to be established. In this way local or readily available plants can be established more rapidly and the asbestos fibers are better sheltered from surface erosion by wind.

5.5 Summary of Plant response in asbestos rich environments

Asbestos rich soils have given rise to some of the most unusual plant communities and some of the most abrupt natural plant boundaries in the world. Vegetation on such sites is under stress and this leads to sparse cover, stunted growth, alterations in colour, extensive root development and species selection. Only selective plants can tolerate the unusual conditions and many serpentine endemic plants have been identified. The diversity of serpentine endemic species is far greater in the unglaciated tropical part of the world presumably because the plants had a much longer time to evolve and adjust to the unusual and toxic conditions. There is often a great contrast between plants growing on and adjacent to serpentine rich sites. Some species grow in both areas but appear to have developed different response mechanisms. Some plants have developed a mechanism for excluding entry of toxic metals into the plant system, while others facilitate the entry and as in the case with nickel translocate the metals rapidly through the plant system. In the latter case key pathways and metabolically sensitive sites are avoided and the excess nickel is stored in the leaves. Citric, malic, and malonic acids seem to play a key role in this process. Plants which have developed such a mechanism are known as hyper-accumulators and as shown by Brooks (1987) some 144 species of plants are capable of storing nickel at concentrations greater than 1000 ug/g nickel. Almost all of them occur in warm temperate to tropical parts of the world.

Attempts to vegetate asbestos rich sites with commercial crops has invariably met with mixed success. Some forbs and grasses could be established and maintained on asbestos rich mine tailings as long as nutrient input was maintained. Other crops germinated but could not be sustained. Tree seedlings invariably have the greatest difficulty and usually only a few survive the first year after planting on such tailings. Selection of plants for revegetation of asbestos waste sites is difficult since no asbestos tolerant commercial cultivars are available. Almost all crops show stress symptoms on such materials and continuous additions of fertilizers to the soils are necessary in order to maintain continuous cover.

The planting of endemic species is desirable but difficult to obtain and the growth of such plants is slow. The same serpentine endemic plants from sites a long distance apart from one another have been shown to develop very different response mechanisms. This is particularly the case when fertilizers are added.

The application of a thick soil cover, and additions of calcium and organic matter seem the most expedient solution and while such modifications are often very expensive they are essential to reduce the hazards of asbestos fiber suspension in the air.

Finally, plants grown on serpentine rich sites are often an undesirable food source for grazing animals since asbestos fibers and trace metals from the soil, and metals accumulated in the plants may be transferred to the animals.

CHAPTER 6

OTHER ENVIRONMENTAL CONCERNS

6.1 Introduction

The presence of asbestos fibers is a concern in many other non-industrial environments. Asbestos fibers are transported by the atmosphere and although concentrations in the outdoor air are generally much lower than in industrial settings, local concentration near asbestos source areas can be significant. Available information on this subject is scarce and a review will be given in the first section of this chapter.

It was previously documented that people living in the vicinity of asbestos rich soils appear to be at some health risk, particularly if they are involved in cultivating and working land containing fibers. A similar concern has been raised with regard to vehicular traffic on this land, including all kinds of off-road vehicular transport such as commercial, military and recreational. The problem might be aggravated by rock quarrying on sites where local bedrock contains significant portions of asbestos fibers, and where bedrock and waste-rock containing asbestos have been used as roadfill and road surface material. River systems draining asbestos rich bedrock environments tend to accumulate sediments rich in asbestos fibers and dredging of such material is problematic because of the redistribution of fibers into the airborne and terrestrial environment.

Finally, asbestos waste disposal will be discussed covering such aspects as removal of asbestos rich sediments in rivers and canals, modifications of waste and deposition of asbestos rich industrial wastes in landfills and reclamation.

Relatively little has been written on these topics and considering the relative durability of many asbestos fibers, it is important to address the potential risk associated with land use options in asbestos rich environments.

6.2 Airborne asbestos fibers in the natural environment

The global distribution of asbestos fibers does not appear to have changed over historic time in spite of significant increases in the use of asbestos. Spurny et al. (1979) claim that on a global basis natural emissions might be greater than those from industrial activities. Local airborne distribution near mining and industrial sites, and in natural settings where soils and sediments are rich in asbestos fibers, may be of greater concern. The mobility of asbestos fibers in the air is aided by the fact that many asbestos rich areas are devoid of, or sparsely covered by, vegetation. It is not easy to state what the actual fiber concentrations are because very few measurements have been made. Since many asbestos rich sites are barren and have major nutrient deficiencies they have not been particularly popular sites for habitation, recreation, and cultivation, hence they have largely been ignored.

With increasing pressure on the land many asbestos rich sites are developed and the potential health ramifications are ignored due to lack of knowledge, insufficient understanding of the potential problem, the tenuous medical evidence, and the long latency period between fiber inhalation and the development of health problems.

Airborne asbestos fibers have been studied primarily in industrial settings. Gibbs and Hwang (1980) have shown distinct differences between fiber type in industrial air. Amosite fibers were consistently longer than crocidolite and chrysotile, and processing of fibers resulted in a reduction of the shorter fiber fraction in the samples. According to the same authors, crocidolite had consistently larger length to width ratios (11.7 -13.2) than amosite (8.4-8.8) and chrysotile (6.2-8.0). In examining the fiber dimensions it is more important to know the fiber type rather than the industrial processing activity since the former appears to have a greater impact on fiber size segregation.

When dealing with airborne fibers the following four components are of major interest: dose, dimension, durability and type of fibers. As shown by Wagner (1986) it is critical to clearly identify the type of fiber present in the air since tremolite and in some cases erionite (zeolite) appear to be a greater threat to health than crocidolite and chrysotile (Dunnigan 1988). Only the combination of electron microscopy and energy dispersive X-ray analysis can be used to differentiate fiber type and, even with this type of technology, distinction between fiber types is often difficult and tedious (Chatfield 1986). The method has been improved using computerization, and semi-automated image analysis techniques ares used extensively to determine fiber burden in lung tissue in cases of Mesothelioma. This provides important information regarding exposure but as shown by McDonald (1988) such data cannot directly be used to determine exposure levels since time related information is needed, particularly as amphibole fibers are retained more rapidly than chrysotile fibers. Also, lung tissue analysis can only be performed after death and thus is not suitable for airborne monitoring purposes. Much improvement of the analytical technique is required before we can consider analyzing air samples for asbestos fiber type on a routine basis. In addition further research is also needed to provide more compelling evidence on the topic of fiber type particularly since some of the man made fibers are also implicated with deteriorating health (Leineweber 1980).

6.2.1 <u>Exposure levels and fiber geometry</u>

Dose is defined as concentration of asbestos over a set duration of time. Given the great variability in the atmosphere this means that many samples must be collected and that a very large number of fibers have to be counted using the Transmission Electron Microscope (TEM). Recent automation and computerization of such analysis has significantly improved the analytical task but the process remains labour intensive and expensive. The main controversy has been over the establishment of threshold exposure levels. In industrial environments most countries have established threshold levels of 2 fibers/liter which assumes that below this threshold the fiber exposure is no longer harmful. The dose exposure curve most applicable to this theory is provided by the curvilinear line (A) in Figure 18. In contrast, a straight linear exposure model (line B, in Figure 18) has also been proposed. The latter case implies that fiber exposure is harmful at all levels and to determine threshold levels becomes very arbitrary. While each theory has its supporters it appears at least in the case of chrysotile that some threshold in exposure level is justified (Dunnigan 1986, Asbestos Institute 1988). However, the threshold concept is considerably more complex because it appears that we have to establish not only a dose threshold, but also a fiber length and width threshold (Stanton et al. 1977,1981, Stanton and Layard 1978) and more recently a fiber durability threshold (Pott 1987).

112

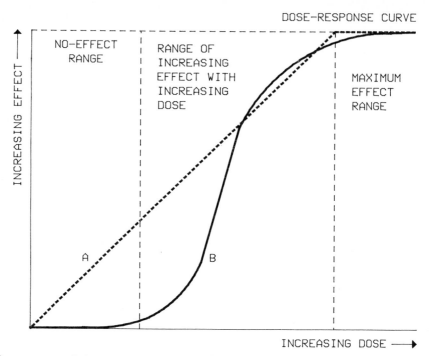

Fig. 15. Models used to determine threshold levels for asbestos fiber exposure (dose) and health effects. A = Continuous exposure without obvious threshold, B = No effect at low level of exposure with obvious threshold level.

Knowledge of fiber dimension is critical since many experiments have shown that long thin fibers are far more hazardous than other fiber configurations. Stanton et al.(1977), Pott (1978) and Stanton et al. (1981) have come up with a model for determining the fiber dimensions which are biologically active. Initially all fibrous particles with length to width ratios >3:1 were thought to be dangerous to human health but Pott (1987) recently recommended to restrict the ratio to >5:1 and considers only fibers longer than 3 um and thinner than 1 um as harmful.

The suspension and deposition efficiency of fibers is dependent on their aerodynamic properties (diameter and length), diffusion coefficients and rate of diffusion charging. Airborne particles have polydispersed size distribution (Cheng 1986) and often clump together (Gentry 1987). Significant efforts have been made to create aerosols of asbestos fibers in the laboratory in order to study the processes of air suspension and distribution (Gentry 1987) and it now appears that the dominant fiber size in the air is below 3um in length.

6.2.2 Durability and fiber type

Finally the durability of the fiber is also of importance and is dependent on the length of fiber contact with body fluid or tissue (Jaurand et al. 1988), type of fluid and fiber chemistry. The persistence of fibers in human tissue is considered most significant in the development of cancer and it is believed that a fiber needs to remain in the tissue until the induction of a tumor. The length of such processes can be decades and this implies that the fiber must be resistant to removal and disintegration. It has been shown in previous chapters that chrysotile is less acid resistant than some of the amphibole fibers and analysis of lung tissue burden have indicated that tremolite is a particularly durable fiber (McDonald 1988, Dunnigan 1988). A durability threshold of more than three years has been proposed by Pott (1987). It is now apparent that different asbestos fibers have different potency and crocidolite, anthophyllite and tremolite appear to be more closely implicated with malignant tumor development that chrysotile fibers. The differences between fiber type and the impurities at the surface of the dust particles can also have a significant impact on tissue toxicity (Le Bouffant 1980).

6.2.3 Measurements in urban and rural settings

While the understanding of the airborne process is improving, actual field measurements in non-occupational settings are still rare. Table 17 summarizes data from some of the published studies and indicates that concentrations by weight and by numbers vary greatly from site to site. These levels are significantly lower than past airborne concentrations measured in indoor industrial settings (Holt 1988) but the measurements are unlikely to be representative of conditions near natural asbestos source areas. Airborne data for such an environment is so fragmentary and scarce that it is impossible to draw any conclusions as to possible exposure levels. What is clear is that urban air is generally higher in fiber concentration than rural air due to the higher concentration of industrial emission sources and larger vehicle traffic releasing fibers from brake linings into the atmosphere (Nicholson et al. 1980, Doll 1987). While it is true that concentration in natural settings is generally low there are indications that levels at specific locations can exceed industrial safety standard levels of 2 fibers / liter (Toft and Meek 1986). Also, high concentration sampling of airborne asbestos is usually carried out using air filters over fixed time intervals and so far no satisfactory analytical model exists to determine fibrous particles on air filters (Spurny 1986). Alternative methods such as collection of particulate matter from the roads and collections via storm sewers have been analyzed to arrive at an indirect indication of airborne emissions (Pitt 1988).

TABLE 17

Airborne asbestos fiber concentrations in the non-industrial environment.

Unaffected emission sources:

Concentration:			Source:	
	< 1	ng/m^3	Source:	Thompson (1978)
	< 10	"		Bruckman (1979)
	10-70	"		IARC (1977)
	0.1-10	"		Charlebois (1978)
	1-5	"		Burdett et al. (1984)
	< 20	"		Nicholson (1984)
	0.1-15	"		Sebastien et al.(1975)
	2-65	"		Selikoff & Lee (1979)
	$1 - 4.5 \times 10^4$	$Fibers/m^3$		Toft and Meek (1986)

Near roads affected by vehicular breaking:

Concentration:			Source:	
	10-25	ng/m^3	Source:	Bruckman (1978)
	2	"		Burdett et al. (1984)
	5×10^4	$Fibers/m^3$		Alste et al. (1976)
	1×10^4	" "		Marfels et al. (1987)

Near industrial emmission sources:

Concentration:			Source:	
	> 30	ng/m^3	Source:	Bruckman (1978)
	28-9700	"		Thompson (1978)
	45- 100	"		Nicholson (1971)
	30-2200	"		Burdett et al. (1984)
	4.7×10^4	$fibers/m^3$		Toft and Meek (1986)
	$<2.6 \times 10^5$	"		Shugar (1979)
	$<9.3 \times 10^5$	"		Shugar (1979)

The occurrence of mesothelioma in rural settings rich in natural asbestos sources as reported by Wagner (1986) is of increasing concern and new research efforts are needed to quantify the emission rates and fiber configurations in such environments. This is of particular importance since fibers other than asbestos might also be implicated (Pelnar 1988).

6.2.4 Airborne emissions near asbestos mines

A topic of particular concern is fugitive dust emissions near asbestos mine sites and asbestos diposal areas. We know that dust generated near asbestos waste sites and mine operations can have a significant impact on the surrounding vegetation and soils. Reductions and changes in soil microbial activities have been reported by Bordeleau et al. (1977) near mining activities, and alteration in soil genesis due to airborne input of asbestos dust has been reported by De Kimpe et al. (1973). It is also of concern to the rural population.

A review of fugitive dust emission from the asbestos mine operations has been provided by Jolicoeur et al. (1984). The major asbestos source in the air comes from the mine extraction process which includes drilling, blasting, transport and disposal of waste in waste-rock piles and tailings. A subsequent source of airborne asbestos comes from weathering and erosion of the waste-rock stockpiles and tailings. The processes of crushing, screening, drying and bagging are carried out in contained areas and are thus subject to the emission control regulation which is generally 2 fibers/cm^3 in the air, standards which are applied in most countries. While there is considerable information on the asbestos emission rates from the contained area of mine operations, there is a big gap in the information regarding asbestos dust emission from drilling, blasting, transporting ore and waste materials.

Some estimates of fugitive dust emissions of 10,000 tons/year from the Quebec asbestos mines have been reported but since the claims were not based on field measurements Jolicoeur et al. (1984) suggest that field measurements are urgently needed before we can produce reliable data. Brulotte (1976) measured dust fall rates in the heart of the Quebec asbestos mine area and came up with rates ranging from 4 - 30 tons/km^2/month, the largest values being obtained closest to the mine waste source area. Laamanen et al. (1965) reported detectable emissions of asbestos fibers some 50 km away from an asbestos mine in Finland. These observations are dependent on wind speed, turbulence and many other factors but they do indicate considerable dispersion of asbestos fibers in the vicinity of asbestos mine waste. All we know for sure is that there are major emissions from the above mentioned mine activities and waste piles and the first step is to reduce emission rates by wetting and other dust reducing methods. A more elaborate discussion on this topic is provided under the section of mine reclamation later in this chapter.

6.3 Vehicular traffic on asbestos bearing material, quarrying, loading and transporting of such materials as part of engineering work

Many rock formations, soils, and sediments contain non-industrial grade asbestos fibers and we have used such materials for road fill, foundations, railroad ballast and other engineering purposes. Some of the materials were used without the knowledge that they contain asbestos and in some cases they were used in foundation and road construction as a means of disposing of waste material. Road and off-road transport on such materials is of considerable concern since it involves the generation of asbestos dust which can reach very hazardous levels.

6.3.1 Asbestos and off-road vehicular traffic

The Clear Creek area in San Benito county in California is underlain by the New Idria-San Benito serpentinite which has extremely high concentrations of chrysotile asbestos. It is a very large federally owned recreation area 185 km southwest of San Francisco. Because of its proximity to a large urban area, the open and hilly terrain has attracted a large segment of society to off-road vehicle (OVR) recreation activities. Popendorf and Wenk (1983) have reported between 20,000 - 40,000 visitor days per year and a large proportion of visitors are involved in off-road motorcycling activities. Because of the dry climate and sparse vegetation the generation of dust from off-road vehicular traffic is of significant concern in view of the large chrysotile component in the surface materials. Cooper et al. (1979) and Popendorf and Wenk (1983) have collected air samples on vehicles and along the road and have reported airborne fiber concentrations between 0.3-5.3 fibers/ml. 90% of the dust collected was chrysotile asbestos and dust accumulation of up to 55 g/m^2/month was recorded along the road access to the recreational area. It was estimated by Cooper et al.(1979) that concentrations in the vicinity of the road were 10^2 to 10^7 times greater than concentrations observed in the urban air.

Efforts have been made to reduce the generation of dust and signs have been posted to warn recreational users of the potential hazards associated with OVR activities on this terrain.

The issue of asbestos dust emission from off-road mobility has also been raised in the US-Army tank training area at Fort Knox, where significant airborne dust levels have been reported (Kruse et al. 1974).

6.3.2 Quarrying, road construction, transportation and dredging

Many rocks contain low grade asbestos fibers and have been quarried and used for road construction, paving and foundations. In Maryland, Rohl et al. (1977) and Carter (1977) have reported several cases where airborne asbestos levels were up to 1000 times larger than those found in unaffected environments. Chrysotile and tremolite fibers were found to be present in the rocks and soils and as a result of quarrying and transportation these materials were suspended in the air. Paving of roads with serpentine material resulted in threatening levels of dust generation from vehicular traffic.

Floodwaters draining asbestos rich areas are known to deposit fibers as part of the sediment load in downstream locations. Schreier (1987) reported a case study on the Sumas river in British Columbia where several fields were inundated with asbestos rich sediments during a flood. The revegetation of such sites is problematic and without application of fertilizers and cover materials these sites constitute a health risk to the local population. Because of such events drainage systems become clogged and dredging becomes necessary. This then creates the new problem of how to properly dispose of the dredged material. Landslides are a frequent phenomenon on serpentinitic bedrock and this factor further contributes to the production and deposition of asbestos rich sediments in the river systems.

Floodwaters draining serpentine deposits in the California Coast Range were found to carry large quantities of asbestos rich sediments into the California Aqueduct. In order to reduce the asbestos fiber concentration in the water and to reduce siltation, dredging of a 16km section of the aqueduct is being carried out (Jones and McGuire (1987). Over 200,000 m^3 of sediments were dredged out of a 16 km canal section in 1982 and 1984. The authors report that a significant reduction in fiber concentration in the water column was achieved as a result of these operations but they fail to describe how the dredged sediments were disposed of. New sediments are likely to be deposited after every spring freshet and this leads to a continuing problem which, according to Hayward (1984), is not restricted to the California Aqueduct but is present in much of northern California.

Given the recent evidence of low hazards associated with ingested fibers (DHHS Committee, 1987) it would appear that the transport, storage and disposal of such dredged sediments are likely more hazardous than simply leaving them in the water system.

Some concern has also been raised by Germine and Puffer (1981) regarding the presence of asbestos fibers in the bedrock of New Jersey, and its use for building materials. Germine (1986) documented that crushed limestone used for play sand contained tremolite, but after additional investigation Langer and Nolan (1987) suggested that the tremolite was massive and not in an acicular form considered hazardous to health. Other environmental concerns have been raised by Bacon et al. (1986) who illustrated the impact on water quality of using asbestos bearing rock material for railroad ballast.

Some concern has also been raised regarding environmental aspects of asbestos-asphalt. Asbestos has long been used in combination with asphalt in road pavement since it offers major technical advantages such as strength, durability, and flexibility. Assessments have been made to evaluate airborne concentrations near pavements in major cities but the rates measured along US and European roads have proven to be very low. Information released by the Asbestos Institute showed that concentrations were generally lower than 0.007 fiber/liter in California, New York, West Germany and France and these concentrations were not significantly different from environmental measurements in the proximity of road pavements that were constructed without asbestos.

6.4 <u>Asbestos waste disposal and reclamation</u>

Two broad categories of waste need to be considered separately: mine waste and industrial waste. The former is usually deposited in mine waste piles and tailings while the second is applied to landfills. The two create different problems and will be discussed separately.

6.4.1 <u>Mine waste and reclamation</u>

Asbestos mines generate large quantities of waste materials that have to be managed efficiently in order to minimize emission of fibers into the environment. Gabra (1984) noted that for each tonne of asbestos fibers, twenty tonnes of asbestos waste is generated. Up to 500 million tonnes of mine tailings have been produced in the eastern township of Quebec, and according to Jolicoeur et al. (1984) the Quebec mines are producing about 40 million tonnes of asbestos rich waste rock and tailings per year.

Dust emission from all these sources can be reduced by physical-mechanical, chemical, and biological techniques. A combination of techniques is needed in most instances. Effective short term dust suppression can be obtained by spraying chemical binders that form a protective coating over the waste material after drying and curing. Some of these have been successfully applied to reduce dust emissions in the transportation of coal. Such compounds as polyacrylate, polyacrylamide and polyvinylacetate are effective but also very costly. Petroleum resins (Coherex), anionic and cationic surfactants that improve wettability, lignosulfonate, and calcium chloride sprays and bitumen emulsions have all been used effectively to suppress dust in other mine operations. Jolicoeur et al. (1984) list a range of options for treating asbestos mine waste and estimated that it would cost up to $ 1,300,000.- per year for continuous cover of waste rock surfactant wetting and treatment of tailings with chemical binders. In terms of long term environmental effects reclamation and revegetation appear to be the ultimate solution, but as shown in Chapters 5 and 6 the establishment of a permanent and sustainable vegetation cover is a complex and expensive proposition. Without the additions of soil, organic rich waste and fertilizers, such revegetation will be difficult. As shown by Jolicoeur et al. (1984) only about 10% of the mine area of Canadian asbestos mines have vegetation cover and it is likely that the situation is similar in other countries. It obviously shows that we have a large and difficult task ahead.

6.4.2 Urban and industrial asbestos waste disposal

There are a number of published guidelines regarding asbestos waste management and disposal (EPA 1985, Cook and Smith 1979). The most common procedure is to apply the waste to trench type landfills and apply a minimum of one to two meters of fill cover over the disposed material to avoid disturbance by vehicles. Because of the enormous quantity of asbestos waste, land fill disposal seems to be the easiest and most expedient approach. Sheils (1984) estimated that up to 120,000 tonnes of asbestos waste are produced on an annual basis in Britain, and Holt (1988) mentioned that up to 100,000 tonnes of asbestos waste have been deposited in one landfill site alone. Some 500 landfill sites have been reported as having received asbestos waste in the UK and according to Baldwin and Heasman (1986) the supply of waste has stayed relatively constant since the reduction in production waste has been matched by a similar increase in demolition waste. The situation in other countries is likely similar.

The waste can be in many forms and as shown by Patel-Mandlik et al. (1988) more than half of the sewage sludge from small cities in New York State contained between 1-5% asbestos. Some urban sewage is dried and bagged and then used as soil amendment in recreation areas and lawns and if it contains asbestos this presents a new health risk particularly for children.

There are many problems associated with asbestos disposal methods. Even when the waste is bagged at the source the deposition process provides some of the highest levels of asbestos fibers in the non-occupational environment (Baldwin and Heasman 1986). Asbestos waste presents a particularly difficult disposal problem since it is generally considered hazardous and non-degradable. As a result, future land use activities on such landfill sites create a considerable risk, particularly since landfills have a tendency to outlast human memories.

Alternatives to direct landfill disposal methods have been investigated by researchers concerned with waste disposal. Suggested alternatives to direct landfill are: compaction, physical, chemical, and thermal modifications and fixations. Given the technical problems and economic consideration progress has been slow. Solidifying asbestos waste by incorporating it into cement or sodium silicate matrices is practised at some sites in the UK (Shiels 1984). A process referred to as "Vitrifix Conversion" also appears to have some potential. In this process the asbestos waste is combined with glass and heated to 1400° C for 10-12 hours to produce a dark brown glass mixture which has a reduced volume and contains up to 80% asbestos (Sheils 1984). Incineration is also being examined for some of the amphibole asbestos wastes.

Asbestos cement waste is thought to be less hazardous and calcium atoms which appear to be scattered over the fiber surface are claimed to contribute to the reduced biological passivity.

Recent evidence presented by Morgan et al. (1977), Monchaux et al. (1981), Jaurand et al. (1984,1988), and Harvey et al. (1984) indicates that malignancy in test animals was significantly reduced when chrysotile asbestos fibers were leached with acids prior to administration to animals. Researchers concerned with waste disposal have followed this lead and experimented with methods to chemically alter the asbestos fiber waste. Mineral waste acids and organic acids created in landfill situations were used to leach chrysotile waste. Baldwin and Heasman (1986) had considerable success in removing up to 25% of the magnesium from the chrysotile material with mineral and organic acids. While this process works reasonably well in the laboratory, field applications are considerably more complex. Type of acid, fiber:fluid ratios, and sample agitation all play important roles in the leaching process. All mineral acids proved to be effective in removing magnesium from the chrysotile fibers, but nitric/chromic acids proved to be more effective over short time periods than sulfuric acid. Artificial and natural mixtures of landfill leachate made up primarily of mixtures of carboxylic acids removed up to 29% of the magnesium from the chrysotile waste after 116 days of exposure.

Several aspects of these experiments which have not been mentioned are: release of trace metals such as nickel, chromium and cobalt, and resistance of some amphibole asbestos fibers to acid leaching. The release of trace metals from asbestos by organic acids has been demonstrated in a natural setting by Schreier (1987) and Schreier et al. (1987a). In addition, if the claims by Wagner (1986) and Dunnigan (1988) are correct, then the alteration of tremolite asbestos fibers is more critical than chrysotile fibers. In contrast to chrysotile, tremolite fibers appear to be very resistant to acid leaching and organic acids such as oxalic acids which readily attach chrysotile (Verlinden et al. 1984) have no effect on tremolite fiber disintegration (Mast and Drever 1987). The disposal of such material is obviously more complex.

A novel approach has been suggested by Gabra (1984); that is to convert chrysotile into magnesium oxide by leaching and solvent extraction with sulfur dioxide. Using SO_2 gas emissions from industrial sources, Gabra claims that extractions of magnesium, iron and nickel are possible using a simple hydrometallurgical process. While this method shows some promise in modifying chrysotile asbestos waste it does not affect some of the more resistant and also more hazardous amphibole asbestos fibers.

A number of other chemical modifications have also been attempted. These are not aimed at fiber dissolution but at modifying the surface physio-chemical properties to passivate the fibers. As noted by Cossette et al. (1986) the coating of fibers with silane and polyphosphates have shown some promise. In the latter process asbestos fibers are subjected to phosphorus oxychloride (Khorami and Nadeau 1986, and Khorami et al. 1987). The authors claim that the surfaces of the asbestos fibers become coated by polyphosphorus without compromising the physical qualities of fibers. However, heat treatment is needed to form insoluble polyphosphate salts which presumably form the protective coat. According to Hodgson (1986) this process is accompanied by a reversal of the zeta-potential of the fibers and thus alters the physio-chemical reactions of such fibers. This method is obviously aimed at industrial applications but might also be of interest in asbestos waste disposal. Unfortunately results so far have been less promising with amphibole fibers.

Finally, the question of concentrating waste in one area or dispersing the hazards by disposing of the asbestos waste in many different sites is a problem that needs further attention. Unfortunately short term economic rather than long term environmental considerations appear to determine the disposal process. While considerable progress is being made in finding alternative methods of asbestos waste disposal it is clear that we have a long way to go in addressing the massive waste disposal problems created by the extensive use of asbestos fibers in recent historic times.

6.5 <u>Summary relating to special environmental concerns</u>

The effects of airborne asbestos fiber exposure in the non-industrial environment have not received much attention since it has generally been believed that exposure levels in the natural environment are usually far lower than in industrial settings. Since fiber inhalation remains a most serious threat to human health it is essential to identify local conditions where bedrock contains high asbestos fiber components. There is insufficient data on airborne concentrations in such environments but the results from off-road vehicle transport studies in a California recreation area clearly illustrate the problem. Similar examples are provided in other parts of the world where such materials are used in industrial activities. The subject of asbestos dust emission from asbestos mine waste rocks and tailings is of particular importance locally since the dust source could potentially be very high. The lack of reliable data in such settings precludes any assessment of the problem and efforts in this regard are urgently needed.

Sediments from asbestos bearing rock formations clog drainage systems, and dredging introduces these materials back into the terrestrial environment. The problem is most acute in central and northern California were the sediment load in the waterway and central distribution system is massive. The removal and disposal of such large quantities of asbestos rich sediments cause enormous difficulties.

The reclamation of asbestos mine waste is a special problem and the use of surfactans and chemical binders helps reduce dust emissions but permanent revegetation appears to be the only appropriate long term solution. Because asbestos has many different nutrient deficiencies and excess magnesium, nickel, cobalt and chromium, revegetation is difficult.

Probably the biggest concern relates to asbestos waste disposal methods in the environmental settings since asbestos is resistant to disintegration and very large quantities of waste have to be disposed of. Gronow (1987) suggests that the rate and degree of dissolution of chrysotile fibers in groundwater in land fills is so slow that there is little likelihood of reducing the pollution in natural waters. Most waste is being deposited in landfills but attempts are being made to chemically modify chrysotile waste by leaching it with mineral and organic waste acids and waste emission gasses. While this will modify the fiber surface in chrysotile, it will not affect amphibole fibers and leaching will also result in trace metal release. However, such processes have only recently been initiated and appear to be only partial solutions to a much larger problem.

EPILOGUE

In assembling data for this review it became apparent that many aspects of asbestos fibers have not been investigated in depth. The subject is a formidable one since asbestos is present to some extent in almost all aspects of the environment. A multi-disciplinary effort is needed to bring the subject into better focus and it is hoped that this monograph will stimulate research interests into aspects of the environment which have previously been neglected.

Because asbestos has been implicated with a unique form of cancer medical research is massive and significant progress in understanding relationships between exposure and health effects have been made. Amongst recent findings the following are significant. Chrysotile fibers are less harmful than amphibole fibers. Low level exposures are less hazardous than high level exposures and at least for chrysotile a threshold "no-effect" level of exposure has been proposed. Limits on fiber geometry (> 5:1 aspect ratios, > 5 um length and < 1um in width) are gaining acceptance and fiber durability thresholds are also proposed. There is some limited evidence of physio-chemical interactions but the question still remains controversial. Finally, the ingestion of high levels of asbestos fibers via drinking water is no longer considered a cause of gastrointestinal cancer.

Asbestos fiber analysis in lung tissue has helped significantly to gain a better understanding of possible causes of tumor development but there are many uncertainties associated with such analysis and they should not be used in isolation or at the expense of other techniques if we hope to make progress in elucidating the mechanism of cancer development.

While these are constructive findings it should be made clear that we have a very challenging road ahead. The implication of amphiboles as a major agent of cancer is posing a considerably more difficult problem. These are among the most complex and variable minerals in nature and the analytical identification techniques are more demanding than those developed for chrysotile. These fibers are also more durable and less easily modified chemically. Hence reduction in biological activity is more difficult to accomplish.

The implication of amphibole fibers with health problems in rural settings in Turkey, Greece and Finland has opened a new challenge to medical researchers. To deal with such problems more interdisciplinary investigations are needed. Not only do we need to know which deposits and soils are hazardous but where they are located, what the mineral formations and weathering products are, and how we can reduce exposure levels.

Although asbestos fibers in water supplies are no longer considered a health threat there are many water issues that need to be examined. Are asbestos fibers affecting fish and aquatic organisms? There is far too little information available to draw any enlightened conclusions on this topic and given the potential of using fish as a cancer test agent it is somewhat surprising that not more effort has been made in this direction. The transport of asbestos fibers in sediments is another area of fruitful research. These sediments are deposited in channels, on land and in the ocean. What is the fate and weathering product of such material? How do we reclaim inundated sites and how do we dispose of such sediments which need to be removed once they clog the river and canal systems. The California Aquaduct water distribution system is one of the obvious examples of such a problem.

Asbestos rich soils cause some of the most unusual boundaries of plant distribution in the world and although much research has been carried out at such sites we are a long way from understanding the causes of poor plant growth. The development of botanical islands consisting of mainly endemic plant communities and the establishments of hyper-accumulating plants are just two examples that are particularly interesting. Unlike the medical researchers, most plant ecologists have focused on chemical effects and what puzzled me is that virtually no investigations have been made to examine what effect (if any) asbestos fibers have on plant root development. Do plants simply avoid fibers and is the abrasive nature of the fibers one of the causes of poor plant growth, or do fibers lodge or even penetrate root cell walls? Some of these are very enticing and challenging research topics.

Evidence exists that biological and microbial activity on such materials is drastically reduced and such organisms as earthworms have great difficulty surviving in such materials. What are the chemical and physical properties that produce such an effect?

Finally, there are two additional topics that need more attention. Reclamation of asbestos mine tailings and disposal of asbestos rich industrial waste. In these cases the concept of : "the best solution to pollution is dilution" does not apply since such fibers are very resistant to dilution and disintegration. The mine tailing reclamation has been a particularly difficult task since asbestos tailings produce a very undesirable medium for plant growth. If no reclamation is done such sites are an excellent source of air pollution and so far planting grasses with massive input of fertilizers has not resulted in the establishment of a sustainable self perpetuated vegetation cover. Similarly, asbestos waste disposal is a problem in many areas and applications to landfills is usually the most expedient short term solution. Unfortunately such waste disintegrates slowly and long term land use plans need to be developed to reduce future hazards. Waste modification with waste acids is being implemented in some parts of the world, but given the resistance of amphibole fibers to such treatment this only solves part of the problem.

There is no doubt that we will gain a better understanding of the many causes and effects of asbestos fibers in the environment in the future. What makes these minerals so fascinating is that they have been enormously useful to man, yet they represent one of the few natural substances that have many adverse effects on man, biota and plants. To expedite our understanding into the mystery of asbestos fibers in the natural environment we need to take a more holistic approach and we need to communicate and cooperate across scientific disciplinary boundaries if we hope to gain a more enlightened picture of this most fascinating natural substance. This monograph does not do justice to these "tiny staws" in wind, water, soil and man but it is a modest start to nourish better cooperation.

128

REFERENCES

Alexander, E.B., 1988. Morphology, fertility and classification of productive soils on serpentinized peridotite in California (U.S.A.). Geoderma, 41: 337-351.

Alexander, E.B., Wildman, W.E. and Lynn, W.C., 1985. Ultramafic serpentinitic) mineralogy class in SSSA: Mineral Classification of Soils, SSSA Spec. Publ. 16, Chapter 12., pp.135-146.

Alste, J., Watson, D. and Bagg, J., 1976. Airborne asbestos in the vicinity of a freeway. Atmospheric Environm. 10: 583-589.

Anderson, A.J., Meyer, D.R. and Mayer, F.K., 1973. Heavy metal toxicities: Levels of nickel, cobalt, and chromium in the soil and plants associated with visual symptoms and variations in growth of an oat crop. Aust. J. Agric. Res., 24: 557-571.

Anderson, A.M., 1902. Historic sketch of the development of legislation for injurious and dangerous industries in England. In: T. Oliver (Editor), Dangerous Trades. Dutton, New York. (7th. edition).

Anderson, C.H. and Long,J.M., 1980. Interim method for determining asbestos in water. US-Environmental Protection Agency, Environ. Res. Laboratory, Athens, Ga., EPA-600/4-80-005, 33 pp.

Antman, K.H., 1986. Asbestos related malignancy. CRC Critical Reviews in Oncology/Hematology, 6: 287-309.

Artvinli, M. and Baris, I., 1979. Malignant mesotheliomas in a small village in the Anatolian Region of Turkey: An epidemiologic study. J. Natl. Cancer Inst. 63 (1): 17-20.

Asbestos Institute, 1988. More scientists now see a threshold in the effect of asbestos. Asbestos, Intern. Bull., 3: 9-10.

Asher, I.M. and McGrath, P.P. (Editors), 1976. Electron microscopy of microfibers; Symposium Proceedings of the First FDA Office of Science Summer Symposium, Penn State University, Aug. 23-25, FDA, Wash., 207 pp.

Atkinson, R.J., 1973. Chrysotile asbestos: colloidal silica surfaces in acidified suspensions. J. Colloid. Interface Sci., 42: 624-628.

Atkinson, A.W. and Rickards, A.L., 1971. Acid decomposition of highly opened chrysotile. Proc. Second Intern. Conf. Physics and Chemistry of Asbestos Minerals. 6-9th Sept. 1971, Louvain Univ., Institute of Natural Sciences, Belgium, 3-1: 10 pp.

Babich, H. and Stotzky, G., 1983. Toxicity of nickel to microbes: Environmental aspects. Adv. Appl. Microbiology, 29: 195-266.

Bacon, D.W., Coomes, O.T., Marsan, A.A. and Rowlands, N., 1986. Assessing potential sources of asbestos fibers in water supplies of S.E. Quebec, Water Res. Bull. 22: 29-38.

Badollet, M.S. and Edgerton, N.W., 1961. The magnetic content of asbestos by magnetic separation. Can. Mining and Metallurgical Bull., 591: 547-550.

Baldwin, G. and Heasman, L.A. 1986. An environmentally acceptable treatment method for chrysotile asbestos waste. In: J.N. Lester (Editor), Proceedings of the International Conference on Chemicals in the Environment. Lisabon, July 1-3. 1986.Selper Ltd., London. pp. 36-46.

Bales, R.C., 1985. Physical and chemical behavior of chrysotile asbestos particles in natural waters. ASCZ Hydraulics Div. Specialty Conf. on : Hydraulics and Hydrology in the Small Computer Age. Orlando, Fl., Aug. 12-17, 1985, 5 pp.

Bales, R.C. and Morgan, J.J., 1985b. Surface charge and adsorption properties of chrysotile asbestos in natural water. Environ. Sci. Technol., 19: 1213-1219.

Barbeau, C., 1979. Evaluation of chrysotile by chemical methods. In: R.L. Ledoux (Editor), Short Course in Mineralogical Techniques of Asbestos Determination. Mineralogical Association of Canada, Quebec, (Section 5), pp. 197-212.

Barbeau, C., Dupuis, M and Roy, J.C., 1985. Metallic elements in crude and milled chrysotile asbestos from Quebec. Environ. Res., 38: 275-282.

Baris, Y.I., 1975. Pleural mesotheliomas and asbestos pleurisies due to environmental asbestos exposure in Turkey: An analysis of 120 cases. Hacettepe Bull. of Med./Surg., 8: 165-185.

Baris, Y., 1980. The clinical and radiological aspects of 185 cases of malignant pleural mesothelioma. In: J.C. Wagner (Editor), Biological Effects of Mineral Fibres. Intern. Agency for Research on Cancer. IARC Scientific Publ. No. 30, WHO, and INAWEM, Lyon, France, 2: 937-947.

Baris, Y.I., Sakin, A.A., Ozesmi, M., Kerse, I., Ozen, E., Kolacan, B., Altinots, M. and Goktepeli, A., 1978. An outbreak of pleural mesothelioma and chronic fibrosing pleurisy in the village of Karain/Urgup in Anatolia. Thorax, 33: 181-192.

Baris, Y.I., Artvinli, M. and Sahin, M.M., 1979. Environmental mesothelioma in Turkey. Ann. N.Y. Acad. Sci., 330: 423-432.

Baris, I., Simonato, L., Artvinli, M., Pooley, F., Saracci, R., Skidmore, J. and Wagner, C., 1987. Epidemiological and environmental evidence of the health effects of exposure to erionite fibers: a four-year study in the Coppadocian Region of Turkey. Intern. Journ. of Cancer, 39: 10-17.

Baron, P.A. and Shulman, S.A., 1987. Evaluation of Magiscan image analyzer for asbestos fiber counting. J. Am. Ind. Hyg. Assoc., 48: 39-46.

Batterman, A.R. and Cook, P.M., 1981. Determination of mineral fiber concentrations in fish tissues. Can. J. Fish. Aquat. Sci., 38: 952-959.

Bazas, T., Bazas, B.,Kitas, D., Gilson, J.C. and McDonald, J.C., 1981. Pleural calcification in North West Greece. Lancet, 1: 254.

Beaman, D.R. and File, D.M., 1976. Quantitative determination of asbestos fiber concentrations. Analytical Chemistry, 48: 101-110.

Beaman, D.R. and Walker, H.J., 1978. Mineral fiber identification using the analytical transmission electron microscope. Proc.of Workshop on asbestos: Definitions and Measurement Methods. N.B.S. Special Publ. 506: 249-270.

Beckett, S.T. and Attfield, M.D., 1974. Inter-laboratory comparisons of the counting of asbestos fiber sampled on membranefibers. Ann. Occup. Hyg., 17: 85-96.

Becklake, M.R., 1982. Asbestos-related fibrosis of the lungs (Asbestosis) and pleura. In: A.P. Fishman (Editor), Pulmonary Diseases and Disorders. McGraw-Hill Book Comp. New York, pp.167-191.

Begin, R., Masse, S., Rola-Pleszczynski, M., Boctor, M. and Drapeau, G., 1988. Asbestos exposure in asbestosworkers and the sheep model. In: G.L. Fisher, and M.G. Gallo (Editors), Asbestos Toxicity. Marcel Dekker Inc., New York/Basel, pp. 87-107.

Belanger, S.E., Cherry, D.S. and Cairns, J., 1986. Uptake of chrysotile asbestos fibers alters growth and reproduction of asiatic clams. Can. J. Fish. Aquat. Sci., 43: 43-52.

Belanger, S.E., Cherry, D.S. and Cairns, J., 1986. Seasonal, behavioral and growth changes of juvenile _corbicula fluminea_ exposed to chrysotile asbestos. Water Res., 20: 1243-1250.

Belanger, S.E., Schurr, K., Allen, D.J. and Gohara, A.F. 1986. Effects of chrysotile asbestos on Coho salmon and green sunfish: evidence of behavioral and pathological stress. Environ. Res., 39: 74-83.

Belanger, S.E., Cherry, D.S., Cairns, J. and McGuire, M.J., 1987. Using asiatic clams as a biomonitor for chrysotile asbestos in public water supplies. J. Am. Water Works Assoc., 79: 69-74.

Bellmann, B., Konig, H., Muhle, H. and Pott, F., 1986. Chemical durability of asbestos and of man-made mineral fibers in vivo. J. Aerosol Sci., 3: 341-345.

Berger, H., 1965. The determination of magnetite in asbestos. Gummi-Asbest-Kunststoffe, 18: 578-582.

Beurlan, H. and Cassedanne, J.P. 1981. The Brasilian mineral resources. Earth Science Review, 17: 177-206.

Biles, B. and Emerson, T.R., 1968. Examination of fibers in beer. Nature, 219: 93.

Bleiman, C. and Mercier, J.P., 1975. Attaque acide et chloration de l'asbeste chrysotile. Bull. Soc. Chim., 3-4: 529-534.

Boatman, E.S., Merrill, T., O'Neill, A., Polissar, L. and Millette, J.R., 1983. Use of quantitative analysis of asbestos fibres in drinking water in the Puget Sound Region. Environ. Health Perspectives, 53: 131-139.

Bolton, R.E., Davis, J.M.G. and Lamb, D., 1982. The pathological effects of prolonged asbestos ingestion in rats. Environ. Res. 29: 134-150.

Bonneau, L., Suquet, H., Malard, C. and Pezerat, H., 1986. Studies on surface properties of asbestos. Environ. Res., 41: 252-267.

Bonser, G.M. and Clayson, D.B., 1967. Feeding of blue asbestos to rats. Br. Emp. Cancer Camp. Res., 45: 242-249.

Bordeleau, L.M., LaLande, R., DeKimpe, C.R., Zizka, J. and Tabi, M. 1977. Effects des poussieres d'amiante sur la microflore tellurique. Plant and Soil, 46:619-627.

Boutin, C., Viallat, J.R., Steinbauer, J., Massey, D.G., Charpin, D. and Mouries, J.C., 1986. Bilateral pleural plaques in Corsica: a non-occupational asbestos exposure marker. Eur. J. Respir. Dis., 69: 4-9.

Brooks, R.R., 1987. Serpentine and its Vegetation. Dioscorides Press Ltd. Portland, Oregon, 454 pp.

Brooks, R.R. and Malaisse, F., 1985. The Heavy Metal Tolerant Flora of South Central Africa - a Multidisciplinary Approach. Balkema,Rotterdam.

Brooks, R.R., Lee, J., Reeves, R.D. and Jaffre, T., 1977. Detection of nickeliferous rocks by analysis of herbarium specimens of indicator plants. J. Geochem. Expl., 7: 49-57.

Brooks, R.R. and Xing-hua Yang, 1984. Elemental levels and relationships in the endemic serpentine flora of the Great Dyke, Zimbabwe and their significance as controlling factors for the flora. Taxon, 33: 392-399.

Brown, A.L., Taylor, W.F. and Carter, R.E., 1976. The reliability of measures of amphibole fiber concentration in water. Environ. Res., 12: 150-160.

Bruckman, L., 1979. A study of airborne asbestos fibers in Connecticut. Workshop on asbestos: Definitions and measurement methods. N.B.S. Spec. Publ. 506 : pp. 179-190.

Brulotte, R., 1976. Study of atmospheric pollution in the Thetford mine area, cradle of Quebec's asbestos industry.Atmos. Pollut. Proc. Int. Colloq., 12: 447-458.

Buelow, R.W., Millette, J.R., McFarren, E. and Symons, J.M., 1980. The behavior of asbestos cement pipe under various water quality conditions: a progress report. J. Am. Water Works Assoc. Feb.,pp 91-102.

Bulusu, K.R., Kulkarn, D.N. and Lutade, S.L., 1978. Phosphate removal by serpentine mineral. Ind. J. Environ. Health, 20: 268-271.

Burdett, G.J., LeGuen, J.M. and Rood, A.P., 1984. Mass concentrations of airborne asbestos in the non-occupational environment, a preliminary report of UK measurements. Ann.Occup. Hyg., 28:31-38.

Burgogne, A.A., 1986. Geology and exploration, McDame asbestos deposit, Cassiar, B.C., CIM Bull., 79: 31-37.

Burlikov, T. and Michailova, L., 1970. Asbestos content of the soil and endemic pleural asbestosis. Environmental Research 3: 443-451.

Butt, B.C., 1981. Exploration forecasts and exploitation realities at the Woodsreef Mine, Nw South Wales, Australia. In: P.H. Riordon (Editor), Geology of Asbestos Deposits, AIME, New York, pp. 63-75.

Carter, L.J., 1977. Asbestos: Trouble in the air from Maryland rock quarry. Science, 197: 237-240.

Case, B. and Sebastien, P., 1987. Environmental and occupational exposures to chrysotile asbestos: A comparative microanalytic study. Arch. Env. Health, 42: 185-191.

Charlebois, C.T., 1978. An overview of the Canadian asbestos problem. Chem. in Canada, pp. 19-38.

Chatfield, E.J., 1979. Measurement of asbestos fibers in the workplace and in the general environment. In: R.L. Ledoux (Editor), Short Course in Mineralogical Techniques of Asbestos Determination. Mineralogical Assoc. Canada, Quebec, (Section 3) pp. 111-163.

Chatfield, E.J., 1986. Asbestos measurements in workplaces and ambient atmospheres. In: S.Basu and J.R. Millette (Editors), Electron Microscopy in Forensic, Occupational, and Environmental Health Sciences. Plenum Press, N.Y. pp. 149-186.

Chatfield, E.J. and Dillon, M.J., 1979. A national survey for asbestos fibers in Canadian drinking water supplies. Environmental Health Directorate, Dept. of Health and Welfare, 79-EHD-34.

Chatfield, E.J., Dillon, M.J. and Stott, W.R., 1983. Asbestos fiber determination in water samples: preparation techniques, improved analytical method, and rapid screening. EPA, US -Environmental Protection Agency, Environ. Res. Lab. Athens, Ga., EPA-600/S4-83-044, 5 pp.

Chatfield, E.J., Dillon, M.J. and Stott, W.R., 1984. Evaluation of turbidimetric methods for monitoring of asbestos fibers in water. Project Summary, EPA-600/S4-84-071.

Cheng, Y.S., 1986. Bivariate lognormal distribution for characterizing asbestos fiber aerosols. Aerosol Sci. Techn., 5: 359-368.

Chidester, A.H., Albee, A.L. and Cady, W.M., 1978. Petrology, structure and genesis of the asbestos-bearing ultramafic rocks of the Belvidere Mountain Area in Vermont, U.S. Geol. Surv. Prof. Paper 1016, 95 pp.

Chisholm, J.E., 1973. Planar defects in fibrous amphiboles, Journ. Material Sci., 8: 475-483.

Chisholm, J.E., 1983. Transmission electron microscopy of asbestos. In: S.S. Chissick and R. Derricott (Editors), Asbestos; Properties, Applications and Hazards. John Wiley & Sons, New York, 2: 85-167.

Chittenden, E.T., Stanton, D.J., Watson, J. and Dodson, K.J., 1967. Serpentine and dunite as magnesium fertilizer. New Zealand J. Agric. Res., 10: 160-171.

132

Choi, I. and Smith, R.W., 1972. Kinetic study of dissolution of asbestos fibers in water. J. Colloid and Interface Sci., 40: 253-261.

Chowdhury, S., 1975. Kinetics of leaching of asbestos minerals at body temperature. J. Appl. Chem. Biotechnol., 25: 347-353.

Chowdhury, S., and Kitchener, J.A., 1975. The Zeta-potentials of natural and synthetic chrysotile. Intern. Journ. Min. Proc., 2: 277-285.

Churg, A., 1988. Chrysotile, tremolite and malignant mesothelioma in man. Chest, 93: 621-628.

Churg, A. and DePaoli, L., 1988. Environmental pleural plaques in residents of a quebec chrysotile mining town. Chest, 94: 58-60.

Clark, P.J., Millette, J.R. and Boone, R.L., 1980. Asbestos-cement products in contact with drinking water: SEM observations. Scanning Electron Microscopy, 1: 341-346.

Cole, M.M., 1973. Geobotanical and biochemical investigations in the sclerophyllous woodland and shrub associations in Eastern Goldfields area of Western Australia. Journ. Appl. Eco. 10: 269-320.

Coleman, R.G., 1977. Ophiolites. Springer, N.Y. 229 pp.

Commins, B.T., 1979. Asbestos in drinking water: a review. Techn. Report TR-100, Water Research Center, U.K., 36 pp.

Commins, B.T., 1983. Asbestos fibres in Drinking Water Commins Associates, Maidenhead, UK., 36 pp.

Commins, B.T., 1984. Asbestos contamination of air, soil and water in perspective. Proc. Intern. Conf. Environmental Contamination. London, U.K., July 1984, UN Envir. Progr. (IRPTC), pp. 21-29.

Commins, B.T., 1985. Ingested Asbestos deemed benign. J.Am. Water Works Assoc., 77: 14-15.

Commins, B.T., 1986. Significance of present-day levels of environmental asbestos. In: J.N. Lester et al. (Editor), Proc. of the Intern. Conf. "Chemicals in the Environment", Lisbon, Portugal, Selper Ltd., London, pp. 484-491.

Commins, B.T. and Gibbs, G.W., 1969. Contaminating organic material in asbestos. Brit. J. Cancer, 23: 358-362.

Committee on Non-Occupational Health Risks of Asbestiform Fibers, 1984. Asbestiform Fibers; Non-occupational Health Risks, National Academy Press; Wash. D.C., 334 pp.

Condie, L.W., 1983. Review of published studies of orally administered asbestos. Environ. Health Perspect., 53: 3-9.

Conforti, P.M., 1983. Effect of population density on results of the study of water supplies in five California counties. Env. Health Perspect., 53: 69-78.

Conforti, P.M., Kanarek, M.S., Jackson, L.A., Cooper, R.C. and Murchio, J.C., 1981. Asbestos in drinking water and cancer incidence in the San Francisco Bay area: 1969-1974. J. Chronic. Dis., 34: 211-224.

Connor, J., Shimp, N.F. and Tedrow, J.C.F., 1957. A spectrographic study of the distribution of trace elements in some podzolic soils. Soil Sci., 83: 65-73.

Constantopoulos, S.H., Goudevenos, J.A., Saratzis, N., Langer, A.M., Selikoff, I.J. and Moutsopoulos, H.M., 1985. Mesovo lung: pleural calcification and restrictive lung function in northwestern Greece. Environmental exposure to mineral fiber as etiology. Environm. Res., 38: 319-331.

Constantopoulos, S.H., Saratzis, N.A., Kontogiannis, D., Karantanas, A., Goudevenos, J.A. and Katsiotgis, P., 1987a. Tremolite whitewashing and pleural calcifications. Chest, 92: 709-712.

Constantopoulos, S.H., Malamou-Mitsi, V.D., Goudevenos, J.A., Papathanasiou, M.P., Pavlidis, N.A. and Papadimitriou, C.S., 1987b. High incidence of malignant pleural mesothelioma in neighbouring villages of Northwestern Greece. Respiration, 51: 266-271.

Cook, J.D. and Smith, E.T., 1979. Dealing with asbestos problems. In: L. Michaels and S.S. Chissick (Editors), Asbestos; Properties, Applications, and Hazards. John Wiley and Sons, N.Y., pp. 279-304.

Cook, P.M., Glass, G.E. and Tucker, J.H., 1974. Asbestiform amphibole minerals: detection and measurement of high concentrations in municipal water supplies. Science, 185: 853-855.

Cook, P.M., Rubin, I., Maggiore, C. and Nicholson, W., 1976. X-Ray diffraction and electron beam analysis of asbestiform minerals in Lake Superior. Inst. Electr. Electron. Eng. J., 75: 34-41.

Cooke, W.E., 1924. Fibrosis of the lungs due to the inhalation of asbestos dust. Br. Med. J. II: 147.

Cooney, P.A., 1987. Environmental asbestos analysis. Analytical Proc., 24: 225-226.

Cooper, R.C. and Murchio, J.C., 1974. Preliminary studies of asbestiform fibers in domestic water supplies. Aerospace Medical Research Laboratory, Wright-Patterson Air Force Base, Ohio, Pap.No. 5: 61-73.

Cooper, W.C., Murchio, J., Popendorf, W. and Wenk, H.R., 1979. Chrysotile asbestos in a California recreational area. Science, 206: 685-688.

Cossette, M., Delvaux, P., VanHa, T., L'Esperance, C. and Belleville, G.G., 1986. Physiological innocuity of asbestos in water. Northeastern Environ. Sci., 5: 54-62.

Cozak, D., Barbeau, C., Gauvin, F., Barry, J.P., DeBlois, C., DeWolf, R. and Kimmerle, F., 1983. The reaction of chrysotile asbestos with titanium (III) chloride. Characterization of the reaction products. Can. J. Chem., 61: 2753-2760.

Craighead,J.E., 1985. Mesothelioma - A plea for biological research. Am. J. Ind. Med., 7: 181-183.

Craighead, J.E., 1988. Response to Dr. Dunnigan's Commentary. Am. J. Ind. Med., 14: 241-243.

Cralley, L.J., Keenan, R.G. and Lynch, J.R., 1967. Exposure to metals in the manufacture of asbestos textile products. Am. Ind. Hyg. Assoc. J., 28: 452-461.

Cralley, L.J., Keenan, R.G., Kujpel, R.E., Kinser, R.E.and Lynch, J.R., 1968. Characterization and solubility of metals associated with asbestos fibers. Am. Ind. Hyg. Assoc. J., 27: 569-573.

Cressey, B.A. and Whittaker, J.W., 1984. Magnetic orientation of amphibole fibres. Can. Mineralogist, 22: 660-674.

Crooke, W.M., 1956. Effect of soil reaction on uptake of nickel from a serpentine soil. Soil Sci., 81:269-275.

Cunningham, H.M. and Pontefract, R., 1971. Asbestos Fibers in Beverages and Drinking Water. Nature. 232:332-333.

Davies, D., 1984. Are all mesotheliomas due to asbestos? Br. Med. Journ., 289: 1164-1165.

Decarreau, A., Colin, F., Herbillon, A., Manceau, A., Nahon, D., Paquet, H., Trauth-Badand, D. and Trescases, J.J., 1987. Domaine segregation in Ni-Fe-Mg Smectites. Clay and Clay Minerals, 35: 1-10.

Decloux, J., Meunier, A. and Velde, B., 1976. Smectite, chlorite, and a regular interlayered chlorite-vermiculite in soils developed on a small serpentinite body - Massive Central, France. Clay Miner., 11: 121-135.

Deer, W.A., Howie, R.A. and Zussman, J., 1962. Rock forming minerals: Vol. 3. Sheet Silicates. Longman, London, pp 170-190.

Deer, W.A., Howie, R.A. and Zussman, J., 1963. Rock Forming Minerals: Vol. 2. Chain Silicates, Longman, London, 379 pp.

DeKimpe, C.R., Tabi, M. and Zizka, J., 1973. Influence of basic mineral on soil genesis in the Thetford-Black Lake area, Province of Quebec. Can. J. Soil Sci., 53: 27-35.

DeWaele, J. K. and Adams, F., 1985. Study of asbestos by laser microprobe. Nato Advanced Sci. Inst. Series. Series B, Physics, 119: 273-274.

DeWaele, J.K., Luys, M.J., Vansant, E.F. and Adams, F.C., 1984. Analysis of chrysotile asbestos by LAMMA and Mossbauer pectroscopy: A study of the distribution of iron. J. Trace and Microprobe Techniques, 2: 87-102.

DeWaele, J.K., Swenters, I.M. and Adams, F.C., 1985. Laser microprobe mass analysis (LAMMA) of organo-silane coated chrysotile fibre surfaces. Spectrochimica acta, Part B. 40 (5-6) 795-800.

DHHS Committee, 1987. Report on cancer risks associated with the ingestion of Asbestos. Environm. Health Perspect., 72: 253-265.

Dixon, J.B., 1977. Kaolinite and serpentine group minerals. In: J.B. Dixon and S.B. Weed (Editors), Minerals in Soil Environments Soil Sci. Soc. Am., Madison, Wisconsin, U.S.A., pp. 357-403.

Doll, R., 1987. The quantitative significance of asbestos fibers in the ambient air. Experientia, 51: 213-219.

Dreyer, C.J.B. and Robinson, H.A. 1981. Occurrence and exploitation of amphibole asbestos in South Africa. In: P.H. Riordon (Editor), Geology of Asbestos Deposits, AIME, New York, pp. 25-44.

Dunnigan, J., 1986. Threshold exposure level for chrysotile. Can. J. of Publ. Health., 77 (1): 41-43.

Dunnigan, J., 1988. Linking chrysotile asbestos with mesothelioma. Am. J. Ind. Med., 14: 205-209.

Durham, R.W. and Pang, T., 1975. Asbestos Fibers in Lake Superior. Special Techn. Publ. 573. American Soc. for Testing and Materials, pp 5-13.

Durham, R.W. and Pang, T., 1976. Asbestiform fibre levels in Lake Superior and Huron. Scientific Series No. 67, Inland Waters Directorate, Environment Canada, Burlington, Ont. 12 pp.

EPA, 1985. Asbestos waste management guidance. EPA, Wash. D.C. EPA/530-SW-85-007.

Ernst, W.H.O., 1976. Physiological and biochemical aspects of metal tolerance. In: T.A. Mansfield (Editor), Effects of air pollutants on plants. Cambridge University Press, Cambridge, pp. 115-133.

Evans, B.W., Johannes, W., Oterdoom, H. and Trommsdorff, V., 1976. Stability of chrysotile and antigorite in the serpentine multisystem. Schweiz. Mineral. Petrogr. Mitt., 56: 79-83.

Fairless, B., 1977. Asbestos fiber concentrations in drinking water of communities using the western arm of Lake Superior as a potable water source. US- Environ. Protect. Agency, EPA-905/4-77-003. Fernandez, E.G. and Martin, F.R., 1986. Comparison of NIOSH and AIA methods for evaluating asbestos fibers: Effects of asbestos type, mounting medium, graticule type and counting rules. Annals of Occup. Hyg., 30: 397-410.

Fisher, G.L., Mossman, B.T., McFarland, A.R. and Hart, R.W., 1987. A possible mechanism of Chrysotile asbestos toxicity. Drug and Chemical Toxicology, 10: 109-131.

Fripiat, J.J. and Faille, M.D., 1966. Surface properties and texture of chrysotiles. Clays and Clay Minerals, 16: 305-320.

Gabra, G., 1984. A process for the production of magnesium oxide from serpentine by sulfur dioxide leaching and solvent extraction. Hydrometallurgy, 13: 1-13.

adsden, J.A., Parker, J. and Smith, W.L., 1970. Determination of asbestos in airborne asbestos by an infrared spectrometer technique. Atmospheric Environ., 4: 667-670.

Gale, R.W. and Timbrell, V., 1980. Practical application of magnetic alignment of mineral fibres for hazard evaluation. In: J.C. Wagner (Editor), Biological Effects of Mineral Fibres. Intern. Agency for Research on Cancer. IARC Sci. Publ. No. 30: 53-60.

Ganotes, J.T. and Tan, H.T., 1980. Asbestos identification by dispersion staining microscopy. Am. Ind. Hyg. Assoc. J., 41: 70-73.

Gaudichet, A., 1978. Asbestos fibers in wines: relation to filtration process. J. Toxicol. Environ. Health, 4: 853-860.

Gentry, J.W., 1987. Survey of recent measurements with asbestos fibers. J. Aerosol Sci., 5: 479-486.

Germine, M., 1986. Asbestos in play sand. New England J. Med., 315: 891.

Germine, M. and Puffer, J.H., 1981. Distribution of asbestos in the bedrock of northern New Jersey area. Environ. Geol. 3: 337-351.

Gibbs, G.W., 1971. The organic geochemistry of chrysotile asbestos from the Eastern Townships, Quebec. Geochimica et Cosmochim. Acta, 35: 485-502.

Gibbs, G.W. and Hwang, C.Y., 1980. Dimensions of airborne asbestos fibres. In: J.C. Wagner (Editor), Biological Effects of Mineral Fibers. IARC Sci. Publ. No. 30: 69-78.

Gibbs, G.W., Baron, P., Beckett, S.T., Dillen, R., Dutoit, R.S.J., Koponen, M. and Robock, K., 1977. A summary of asbestos fiber counting experience in seven countries. Annals of Occup. Hyg., 20: 321-332.

Goni, J., Johan, Z. and Sarcia, C., 1971. Study of the physical and crystallographic properties of chrysotile lixiviated by oxalic acid reaction. Proc. Second Intern. Conf. Physics and Chemistry of Asbestos Minerals. 6-9. September 1971. Louvain University, Institute of Natural Sciences, Louvain, Belgium, Comm., pp. 1-2.

Gravatt, C.C., La Fleur, P.D. and Heinrich, K.F.J.,(Editors), 1978. Proc. of Workshop on Asbestos: Definitions and Measurement Methods. National Bureau of Standards, Gaithersburg, MD, July 18-20. 1977, N.B.S. Special Publ. 506: 496 pp.

Graney, R.J., Cherry, D.S. and Crains, J., 1983. Heavy metal indicator potential of the Asiatic Clam (Corbicula fluminea) in artificial stream systems. Hydrobiologia, 102: 81-85.

Great Lake Research Advisory Board, 1975. Asbestos in the Great Lakes basin with emphasis on Lake Superior. Intern. Joint Commission, USA and Canada, Report No. STIJC-7503 : 51 pp.

Gronow, J.P., 1987. The dissolution of asbestos fibers in water. Clay Minerals, 22: 21-35.

Gross, P., Harley, R.A., Swinburn, L.M. and Davis, J.M.G. and Green, W.B., 1974. Ingested mineral fibers, do they penetrate tissue or cause cancer? Arch. Environ. Health. 29: 341-347.

Haab, R., 1988. Chrom- und Nickelverteilung in Rohhumusauflagen auf Serpentinit. Eidgenossische Technische Hochschule (ETH), Labor fur Bodenkunde, Zurich, Diplomarbeit, ETH, 85 pp.

Hahn-Weinheimer, P. and Hirner, A., 1975. Major and trace elements in Canadian asbestos ore bodies - Analytical results and statistical interpretation. Troisieme conference internationale sur la physique et la chimie des mineraux d'amiante, Aout, Universite Laval, Quebec, 6: 25.

Hahn-Weinheimer, P. and Hirner, A., 1977. Influence of hydrothermal treatment on physical and chemical properties of asbestos. J.& Proc. Royal Soc. New South Wales, 110: 99-110.

Hallenbeck, W.H., Chen, E.H., Patel-Mandlik, K. and Wolff, A.H., 1977. Precision of analysis for waterborne chrysotile asbestos by transmission electron microscopy. Bull. Environ. Contam. Toxic., 17: 551-558.

Hallenbeck, W.H., Chen, E.H., Hesse, C.S., Patel-Mandlik, K. and Wolff, A.H., 1978. Is chrysotile asbestos released from asbestos cement pipe into drinking water? J. Am. Water Works Assoc., Feb., 97-102.

Halstead, R.L., 1968. Effect of different amendments on yield and composition of oats grown on soils derived from serpentine Material. Can. J. Soil Sci., 48: 301-305.

Halstead, R.L., Finn, B.J. and MacLean, A.J., 1969. Extractability of nickel added to soils and its concentration in plants. Can. J. Soil Sci., 48: 301-305.

Hanley, D.S., 1987. The origin of the chrysotile asbestos veins in Southeastern Quebec. Can. J. Earth Sci. 24:1-9.

Harben, P.W. and Bates, R.L., 1984. Asbestos. Geology of the nonmetallics. Metal Bulletin Inc., New York, pp. 324-335.

Harington, J.S., Allison, A.C. and Badami, D.V., 1975. Mineral fibers: Chemical, physicochemical, and biological properties. Adv. in Pharmacology and Chemotherapy, 12: 291-402.

Harrington, J.M., Craun,G.F., Meigs, J.W., Landrigan, P.J., Flannery, J.T. and Woodhull, R.S., 1978. An Investigation of the Use of Asbestos Cement Pipe for Public Water Supply and the Incidence of Gastrointestinal Cancer in Connecticut. 1935-1978. Am. J. Epidemiol. 107:96-103.

Harris, A.M. and Grimshaw, R.W., 1971. The leaching of ground chrysotile. Proc. Second Intern. Conf. Physics and Chemistry of Asbestos Minerals. 6-9th Sept. 1971., Louvain Univ., Institute of Natural Sciences, Belgium, 3-2: 6 pp.

Harris, A.M. and Grimshaw, R.W., 1975. The leaching of ground chrysotile. 3rd International Asbestos Conference, Quebec, 1975.

Harvey, G., Page, M. and Dumas, L., 1984. Binding of Environmental Carcinogens to Asbestos and Mineral Fibres. Br. J. Ind. Med. 41:396-400.

Hayward, S.B., 1984. Field monitoring of chrysotile asbestos in California waters. Jour. AWWA, 76: 66-73.

Health and Welfare Canada, 1979. A national survey for asbestos fibers in Canadian drinking water supplies, Environmental Health Directorate, Health and Welfare Canada, Ottawa, Report 79-34.

Henson, E.B., 1985. Asbestos fibers in lakes and streams. Intern. Assoc. of Theoretical and Applied Limnology Proc., 22: 2232-2239.

Herman, R.L., 1985. Mesothelioma in rainbow trout, Salmo gairdneri Richardson. J. Fish Diseases, 8: 373-376.

Hesse C.S, Hallenbeck, W.H., Chen, E.H. and Brenniman, G.R., 1977. Determination of chrysotile asbestos in rainwater. Atmo. Environ., 11:1233-1237.

Hilding, A.C., Hilding, D.A. Larson, D.A. and Aufdenheide, A.C., 1981. Biological Effects of Ingested Amosite Asbestos, Taconite Tailings, Diatomaceous Earth and Lake Superior Water in Rats. Arch. Environ. Health. 36: 298-303.

Hillerdal, G., 1985. Nonmalignant pleural disease related to asbestos exposure. Clinics in Chest Med., 6: 141-152.

Hodgson, A.A., 1979. Chemistry and physics of asbestos. In: L. Michaels & S.S. Chissick (Editors), Asbestos; Properties, Applications, and Hazards. John Wiley & Sons, New York, 1: 67-114.

Hodgson, A.A., 1986. Scientific advances in asbestos 1967-1985. Anjalena Publ. Crowthorne, Berkshire, U.K., 186 pp.

Holmes, A., Morgan, A. and Scandalls, F.J., 1971. Determination of iron, chromium, cobalt, nickel and scandium in asbestos by neutron activation analysis. Am. Ind. Hyg. Assoc.J., 32 : 281-286.

Holt, P., 1988. Asbestos dust. Chemistry in Britain, September Issue, pp. 903-906.

Horler, D.N.H., Barber, J. and Barringer, A.R., 1981. New concepts for the detection of geochemical stress in plants. In: J.A. Allan and M. Bradshaw (Editors), Remote Sensing in Geology and Terrain Studies, Remote Sensing Society, London, pp 113-123.

Hughes, M.K., 1981. Cycling of trace metals in ecosystems. In: Lebb, N.W. (Editor), Effect of heavy metal pollution on plants. Applied Science Publishers, London/New Jersey, Vol. 2: pp. 95-118

Hughes, M.K., Lepp, N.W. and Phipps, D.A., 1980. Aerial heavy metal pollution and terrestrial ecosystems, Adv. Ecol. Res., 11: 217-227.

Huncharek, M., 1986. The biomedical and epidemiological characteristics of asbestos-related diseases: A review. The Yale J. of Biology and Medicine, 59: 435-451.

Hunsinger, R.B. and Roberts, K.J., 1980. Chrysotile asbestos fiber removal during potable water treatment. Environ. Sci. Technol., 14: 333-336.

Hunter, J.G. and Vergnano, O., 1952. Nickel toxicity in plants. Ann. Appl. Biol., 39: 279-284.

Hutchison, J.L., Irusteta, M.C. and Whittaker, E.J.W., 1975. High-resolution electron microscopy and diffraction studies of fibrous ampiboles. Acta Cryst., A31: 794-801.

IARC Working Group, 1977. Monographs on the evaluation of the carcinogenic risk of chemicals to man: Asbestos. International Agency for Research on Cancer, Lyon, France, Vol. 14, 30pp.

Isherwood, R. and Jennings, B.R., 1985. Electrooptical characteristics of chrysotile asbestos soils. J. Colloid & Interface Sci., 108: 462-470.

Istock, J.D. and Harward, M.E., 1982. Influence of soil moisture on smectite formation in soils derived from serpentinite. Soil Sci. Soc. Am. J., 46:1106-1108.

Jacobasch, H.J., Baubock, G. and Schurz, J. 1985. Polymer science. Problems and results of zeta-potential measurements on fibers. Colloid & Polymer Sci., 263: 3-24.

Jaffre, T., 1980. Etude ecologique du peuplement vegetal des sols derives de roches ultrabasiques en Nouvelle Caledonie. ORSTOM, Paris.

Jarvis, S.C., 1984. The association of cobalt with easily reducible manganese in some acidic permanent grassland soils.J. Soil Sci., 35: 431-438.

Jaurand, M.C., Gaudichet, A., Halpern, S. and Bignon, J., 1984. In vitro biodegradation of chrysotile fibres by alveolar macrophages and mesothelial calls in culture: comparison with a pH effect. Brit. J. Ind. Med., 41: 389-395.

138

Jaurand, M.C., Renier, A., Gaudichet, A., Kheuang, L., Magne, L. and Bignon, J., 1988. Short-term tests for the evaluation of potential Cancer Risk of modified asbestos fibers. Annals: N.Y. Acad. Sci., 534: 741-753.

Jefferson, D.A., Mallinson, L.G., Hutchinson, J.L. and Thomas, J.M., 1978. Multiple chain and other unusual faults in amphiboles. Contrib. Mineralogy and Petrology, 66: 1-4.

Jenkins, D.M., 1987. Synthesis and characterization of tremolite in the system H2O-CaO-MgO-SiO2. Am. Mineralogist, 72: 707-715.

Johnston, W.R. and Proctor, J., 1981. Growth of serpentine and non-serpentine races of Festuca rubra in solutions simulating the chemical conditions in a toxic serpentine soil. J. Ecology, 69: 855-869.

Jolicoeur, C. and Poisson, D., 1987. Surface physio-chemical studies of chrysotile asbestos and related minerals. Drug and Chem. Toxicology, 10: 1-47.

Jolicoeur, C., Kwak, J.C.T. and LeBel, J., 1984. Dust emission in asbestos mining operations and asbestos waste disposal. Environment Canada, Environmental Protection Service, Quebec, and PRAUS, Programme de recherche sur l'amiante de l'Universite de Sherbrooke, Sherbrooke, Quebec, 126 pp.

Jones, J. and McGuire, M.J., 1987. Dredging to reduce asbestos concentrations in the California aqueduct. J. Am. Water Works Assoc., 79: 30-37.

Jones, R.C., Hudnall, W.H. and Sakai, W.S., 1982. Some highly weathered soils in Puerto Rico, 2. Mineralogy. Geoderma 27: 75-138.

Juchler, S. and Sticher, H., 1985. Der Totalpbergsturz bei Davos aus Bodenkundlicher Sicht. Geographica Helvetica, 3: 123-132.

Kanarek, M.S., Conforti, P.M., Jackson, L.A., Cooper, R.C. and Murchio, J.C., 1980. Asbestos in drinking water and cancer incidence in the San Francisco Bay area. Am J. Epidemiol., 112: 54-72.

Kanno, I., Onikura, Y. and Tokudode, S., 1965. Genesis and characteristics of brown forest soils derived from serpentinite in Kyushu, Japan. Soil Sci. and Plant Nutr., 11: 225-234.

Kay, G., 1973. Ontario intensifies search for asbestos in drinking water. Water Pollut. Control Fed., Sept.: 33-35.

Kay, G., 1974. Asbestos in drinking water. J. Am. Water Works Assoc., 66: 513-514.

Khorami, J. and Nadeau, D., 1986. Physicochemical characterization of asbestos and attapulgite mineral fibers before and after treatment with phosphorus oxychloride. Thermochimica Acta., 108: 279-287.

Khorami, J., Lemieux, A. and Nadeau, D., 1987. The phosphorylation of chrysotile asbestos fibers with phosphorus oxychloride (POCl3): mechanism of reaction and chemical composition of the external coating. Can. J. Chemistry. 65: 2268-2276.

Kinzel, H., 1982. Serpentine Pflanzen. In: Pflanzenphysiologie und Mineralstoff Wechsel, Ulmer Stuttgart.

Kirkman, J.H., 1975. Clay minerals in soils derived from ultrabasis rocks of southern Westland, N.Z. N.Z. J. Geol. Geophys. 18: 849-864.

Kiviluoto, R., 1960. Pleural calcification as a roentgenologic sign of non-occupational endemic anthophyllite-asbestosis. Acta Radiol. Ther. Phys. Biol. Suppl., 194: 1-67.

Kiviluoto, R., 1965. Pleural plaques and asbestos. Ann. N.Y. Acad. Sci., 132: 235-239.

Krantz, S., 1987. The asbestos problem from a measuringviewpoint. National Board of Occup. Health Newsletter, 1-2: 8-11.

Krause, W., 1958. Andere Bodenspezialisten. Handbuch der Pflanzenphysiologie, Band VI, Die mineralische Ernahrung der Pflanze, Springer Verlag, Berlin, pp. 755-806.

Kruckeberg, A.R., 1954. The ecology of serpentine soils: A symposium. III Plant species in relation to serpentine soils. Ecology, 35: 267-274.

Kruckeberg, A.R., 1967. Ecotypic response to ultramafic soils by some plant species in northwestern United States. Brittonia, 19: 133-151.

Kruckeberg, A.R., 1969. Plant life on serpentinite and other ferromagnesian rocks in northwestern North America. Syesis, 2: 15-114.

Kruckeberg, A.R., 1979. Plants that grow on serpentine - A hard life. Davidsonia, 10: 21-29.

Kruckeberg, A.R., 1984. California serpentines: Flora, vegetation, geology, soils, and management problems. Univ. of California Press, Berkeley, 180 pp.

Kruse, C.A., Carey, P.H. and Howe, D.J., 1974. Silica content in dust from tank ranges . Army Medical Research Laboratory, Fort Knox, KY. Report # USAMPL-1, 13 pp.

Kuryval, R.J., Wood, R.A. and Barrett, R.E., 1972. Identification and assessment of asbestos emissions from incidental sources of asbestos. US-EPA, 650/2-74-087: 286 pp.

Laamanen, A., Noro, L. and Raunio, V., 1965. Observation on atmospheric air pollution caused by asbestos. Ann. N.Y. Acad. Sci., 132: 240-248.

Lafuma, J., Morin, M., Poncy, J.L., Masse, R., Hirsch, A., Bignon, J. and Monchaux, G., 1980. Mesothelioma induced by intrapleural injection of different types of fibers in rats; synergistic effect of other carcinogens. In: J.C. Wagner (Editor), Intern. Agency for Research on Cancer. IARC Sci. Publ. 30: 311-320.

Lamarche, R.Y. and Riordon, P.H., 1981. Geology and genesis of the chrysotile asbestos deposits of northern Appalacia. In: P.H. Riordon (Editor), Geology of Asbestos Deposits. Society of Mining Engineers, American Inst. Mining, Metallurgical, and Petroleum Engineers, Inc. New York, pp. 11-23.

Langer, A.M. and Nolan, R.P., 1986. Asbestos in potable water supplies and attributable risk of gastrointestinal cancer. Northeastern Environ. Sci., 5: 41-53.

Langer, A.M., Mackler, A.D. and Pooley, F.D., 1974. Electron microscopical investigations of asbestos fibers. Environ. Health Perspect., 9: 64-72.

Langer, A.M., Maggiore, C.M., Nicholson, W.J., Rohl., A.N., Rubin, I.B. and Selikoff, I.J., 1979. The contamination of Lake Superior with amphibole gangue minerals. Ann. N.Y. Acad. Sci., 330: 549-472.

Langer, A.M., Nolan, R.P. and Constantopoulos, H.M., 1987. Association of Metsovo lung and pleural mesothelioma with exposure to tremolite-containing whitewash. Lancet, 1: 965-967.

Lauth, J. and Schurr, K., 1983. Some effects of chrysotile asbestos on a planktonic algae (Cryptomonas Erosa). Micron, 14: 93-94.

Lauth, J. and Schurr, K., 1984. Entry of chrysotile asbestos fibres from water into the planktonic algae (Cryptomonas Erosa). Micron and Microscopica Acta, 15: 113-114.

Lawrence, J. and Zimmermann, H.W., 1976. Portable water treatment for some asbestiform minerals: Optimization and turbidity data. Water Res., 10: 195-198.

140

Lawrence, J. and Zimmermann, H.W., 1977. Asbestos in water: mining and processing effluent treatment. J. Water Pollut. Control Fed., 49: 156-170.

Lea, F.M. 1970. The chemistry of cement and concrete. 3rd edition, Edward Arnolds Ltd. London, U.K. 727 pp.

Le Bouffant, L., 1980. Physics and chemistry of asbestos dust. In: J.C. Wagner (Editor), Biological Effects of Mineral Fibers. IARC Sci. Publ. No. 30, pp. 15-23.

Leclerc, A., Goldberg, M., Goldberg, P., Delqumeaux, J. and Fuhrer, R., 1987. Geographical distribution of respiratory cancer in New Caledonia. Arch. Environ. Health, 42: 315-320.

Lee, D.H.K. and Selikoff, I.J., 1979. Historical background to the asbestos problem. Environ. Research 18: 300-314.

Lee, W.G. and Hewitt, A.E., 1981. Soil changes associated with development of vegetation on an ultramafic scree, northwest Otago, New Zealand. J. of the Royal Society of New Zealand, 12: 229-242.

Lee, J, Reeves, R.D., Brooks, R.R. and Jaffre, T., 1977. Isolation and identification of a citrato-complex of nickel from nickel-accumulating plants. Phytochemistry, 16: 1503-1505.

Lee, J., Reeves, R.D., Brooks, R.R. and Jaffre, T., 1978. The relation between nickel and citric acid in some nickel-accumulating plants. Phytochemistry, 17: 1033-1035.

Leight, W.G. and Wei, E.T., 1977. Surface charge and asbestos toxicity. Nature, 265: 537-539.

Leineweber, J.P., 1978. Statistics and the significance of asbestos fiber analysis. Proc. of Workshop on Asbestos: Definitions and Measurement Methods. N.B.S. Special Publ. 206:281-294.

Leineweber, J.P., 1980. Dust chemistry and physics: Mineral and vitreous fibres. In: J.W. Wagner (Editor), Biological Effects of Mineral Fibres. Intern. Agency for Research on Cancer, Scientific Publ.# 30, WHO & INSERM, Lyon, France, 2: 881-900.

Levine, D.S., 1985. Does asbestos exposure cause gastrointestinal cancer? Digestive Diseases and Sciences, 30: 1189-1198.

Levy, B.S., Sigurdson, E., Mandel, J., Laudon, L. and Pearson, J., 1976. Investigating possible effects of asbestos in City Water Surveillance of Gastrointestinal Cancer Incidence in Duluth, Minnesota. Am. J. Epidemiol., 103: 362-368.

Light, W.G. and Wei, E.T., 1977. Surface charge and asbestos toxicity. Nature, 265: 539-539.

Lin, F.C. and Clemency C.V., 1981. The dissolution of brucite, antigorite, talc and phlogopite at room temperature and pressure. Am. Mineralogist.,66: 801-806.

Lockwood, T.H., 1974. The analysis of asbestos for trace metals. Am. Ind. Hyg. Assoc. J., 5: 224-250.

Lyon, G.L., 1969. Trace elements in New Zealand plants. PhD Thesis, Massey University, Plamerston North, New Zealand.

Lyon, G.L., 1971. Calcium, magnesium and trace elements in a New Zealand serpentine flora. J. Ecol., 59: 421-429.

Lyon, G.L., Brooks, R.R., Peterson, P.J. and Butler, G.W., 1968. Trace elements in a New Zealand serpentine flora. Plant Soil, 29: 225-240.

Lyon, G.L., Peterson, P.J. and Brooks, R.R., 1969. Chromium-51 distribution in tissue and extracts of Leptospermum scoparium. Planta, 88: 282-287.

MacRae, K.D. 1988. Asbestos in Drinking Water and Cancer. Journ. Royal College of Physicians of London, 22: 7-10.

Madhok, O.P. and Walker, R.B., 1969. Magnesium nutrition of two species of sunflower. Pl. Physiol., 44: 1016-1022.

Magee, F., Wright, J.L., Chan, N. Lawson, L. and Churg, A., 1986. Malignant mesothelioma caused by childhood exposure to long-fiber low aspect ratio tremolite. Am. J. of Indust. Med. 9: 529-533.

Main, J.L., 1974. Differential responses to magnesium and calcium by native populations of Agropyron spicatum. Amer. J. Bot., 61: 931-937.

Main, J.L., 1981. Magnesium and calcium nutrition of a serpentine endemic grass. Am. Midl. Nat., 105: 196-199.

Maresca, G.P., Puffer, J.H. and Germine, M., 1984. Asbestos in lake and reservoir water of Staten Island, New York: Source, concentration, mineralogy, and size distribution. Environ. Geol. Water. Sci., 6: 201-210.

Marfels, H., Spurny, K.R., Jaekel, F., Boose, C., Althaus, W., Opiela, H., Schormann, J., Weiss, G. and Wulbeck, F.J., 1987. Asbestos fiber measurements in the vicinity of emitents. J. Aerosol Sci., 18: 627-630.

Marsh, G.M., 1983. Critical review of epidemiologic studies related to ingested asbestos. Environ. Health Perspect., 53: 49-56.

Martilla, R.K. 1979. Fibrous minerals in the Wabush Iron Ore District, Labrador, Canada. MSc thesis, McMaster University, Hamilton, Ont., 130 pp.

Martinez, E. and Zucker, G.L., 1960. Asbestos body materials studied by Zeta potential measurements. J. Phys. Chem., 64: 924-926.

Mason, T.J., McKay, F.W. and Miller, R.W., 1974. Asbestos Like Fibers in Duluth Water Supply: Relation to Cancer Mortality. J. Am. Med. Assoc., 228:1019-1020.

Mast, M.A. and Drever, J.I. 1987. The effect of oxalate on the dissolution rates of oligoclase and tremolite. Geochimica Cosmochimica Acta, 51: 2559-2568.

Matti, M.A. and Al-Adeeb, A., 1985. Sulphate attack on asbestos cement pipes. Intern. J. Cement Composites & Lightweight Concrete, 7: 169-176.

McConnel, E.E., 1984. NTP technical report on the toxicology and carcinogenesis studies of chrysotile asbestos. U.S. Department of Health and Human Services, PHS, NIH, Research Triangle Park, N.C., NIH Publication No. 84-2551.

McConnel, E.E., Shefner, A.M., Rust, J.H., and Moore, J.A., 1983a. Chronic effects of dietary exposure to amosite and chrysotile asbestos in Syrian golden hamsters. Env. Health Perspect., 53: 11-25.p

McConnel, E.E., Rutter, H.A., Ulland, M.B. and More, J.A., 1983b. Chronic effects of dietary exposure to amosite asbestos and tremolite in F 344 rats. Env. Health Perspect., 53: 27-44.

McConnochie, K., Simonato, L. and Wagner, J.C., 1986. Mesothelioma in Cyprus - the role of tremolite. Thorax, 41(3): p. 250.

McConnochie,K., Simonato, L., Mavrider, P., Christofides, P., Pooley, F.D. and Wagner, J.C., 1987. Mesothelioma in Cyprus - The role of tremolitic. Thorax, 42: 342-347.

McCrone, W.C., 1978. Identification of asbestos by polarized light microscopy. Proc. of Workshop on Asbestos: Definitions and Measurement Methods, N.B.S. Special Publ. 509: 235-247.

McDonald, J.C., 1980. Malignant mesothelioma in Quebec. In: J.C. Wagner (Editor) Biological Effects of Mineral Fibers. IARC Sci. Publ. No. 20, Vol. 2:673-680.

McDonald, J.C., 1988. Tremolite and other amphiboles and mesothelioma. Am. J. Ind. Med., 14: 247-249.

McDonald, J.C. and McDonald, A.D., 1977. Epidemiology of mesothelioma from estimated incidence. Prev. Med. 6: 426-446.

McDonald, A.D. and McDonald, J.C., 1980. Malignant mesothelioma in North America. Cancer, 46: 1650-1656.

McGuire, M.J., Bowers, A.E. and Bowers, D.A. 1982. Asbestos analysis case history: surface water supplies in Southern California. Journ. AWWA., 74: 470-477.

McLaren, R.G.. Lawson, D.M. and Swift, R.S., 1986. The forms of cobalt in some Scottish soils as determined by extraction and isotopic exchange. J. Soil Sci., 37: 223-234.

McMillan, L.M., Stout, R.G. and Willey, B.F., 1977. Asbestos in raw and treated water: an electron microscopy study. Environ. Sci. Techn., 11: 390-394.

Meigs, J.W., Walter,S.D., Heston,F.J., Millette, J.R., Craun, G.F., Woodhull,R.S. and Flannery, J.T., 1980. Asbestos Cement Pipe and Cancer in Connecticut 1955-1974. J. Environ.Health., 42: 187-191.

Menezes de Sequeira, E., 1968. Toxicity and movement of heavy metals in serpentine soils (North-Eastern Portugal). Agronomia Lusit., 30: 115-154.

Merewether, E.R.A. and Price, C.W., 1930. Effects of asbestos dust on lungs and data suppression in asbestosis and carcinoma of the lung. In: Annual Report of the Chief Inspector of Factories for the Asbestos Industry. London, H.M. Stationery Office.

Metson, A.J. and Gibson, E., 1977. Magnesium in New Zealand soils. N.Z. Journal of Agricultural Research, 20: 163-184.

Meyer, D.R., 1980. Nutritional problems associated with the establishment of vegetation on tailings from an asbestos mine. Env. Poll. (Series A), 23: 287-298.

Meyer, E., 1982. Untersuchungen zum Vorkommen von Asbestfasern in Trinkwasser in der Bundesrepublik Deutschland und gesungheitliche Bewertung der Ergebnisse. GWF-Wasser/Abwasser, 123: 85-96.

Millette, J.R., 1983. Asbestos in water supplies of the United States, Summary Workshop on Ingested Asbestos. Environ. Health Perspect., 53: 45-48.

Millette, J.R. and Kinman, R.N., 1984. Iron-containing coatings on asbestos-cement pipes exposed to aggressive water. Proc. 11. Am. Water Works Assoc.Conf. Water Quality Techn., 11: 171-184.

Millette, J.R., Clark, P.J. and Pansing, M.F., 1979. Exposure to asbestos from drinking water in the United States. Environmental Health Effects Research Report, Office of Research and Development, EPA-600/1-79-028.

Millette, J.R., Boone, R. and Rosenthal, M., 1980a. Asbestos in cistern water. Environmental Research Brief. US Environmental Protection Agency, Cincinnati, Ohio, February 1980, 4 pp.

Millette, J.R., Clark, P.J., Pansing, M.F. and Twyman, J.D., 1980b. Concentration and size of asbestos in water supplies. Environ. Health Perspect., 34: 13-25.

Millette, J.R.; Boone,R.L., Rosenthal, M.T. and McCabe, L.J., 1981a. The need to control asbestos fibers in potable water supply systems. Sci. Total Env., 18: 91-102.

Millette, J.R., Pansing, M.F. and Boone, R.L., 1981b. Asbestos-cement materials used in water supply. J. Am. Water Works Assoc., March: 48-51.

Millette, J.R., Craun, C.S., Stober, A., Kramer, D.F., Tousignant, H.G., Hildago, E., Duboise, R.L. and Benedict, G., 1983. An epidemiologic study of the use of asbestos-cement pipe for the distribution of drinking water in Escambia County, Florida. Env. Hlth. Perspect., 53: 91-98.

Millette, J.R., Clark, P.J., Boone, R.L. and Rosenthal, M.T., 1987. Occurrence and biological activity testing of particulates in drinking water. Bull. Environ. Contam. Toxicol., 31: 1-8.

Minguzzi, C. and Vergnano, O., 1948. Il contenuto di nichel nelle ceneri di *Alyssum bertolonii* Desv., Atti Soc. Yosc. Sc. Natur., Mem. Pisa, 55: 49-74.

Mizuno, N., 1979. Studies of the chemical composition of serpentine soils and mineral deficiencies and toxicities of crops. Rep. Hokk. Pref. Ag. Exptl. Stns., 29: 1-77.

Monaro, S., Landsberger, S., Lecomte, R. and Paradi, P., Desaulnier, S.G. and P'an, A., 1983. Asbestos pollution levels in river water measured by proton-induced X-ray emission (PIXE) techniques. Environ. Pollut. Ser. B., 5: 83-90.

Monchaux, G., Bignon, J., Jaurand, M.C., Lafuma, J., Sebastien, P., Masse, R.,Hirsch, A. and Goni, J., 1981. Mesotheliomas in rats following ionculation with acid-leached chrysotile asbestos and other mineral fibers. Carcinogenesis, 2: 229-236.

Monkman, L.J., 1971. Some chemical and mineralogical aspects of the acid decomposition of chrysotile. Proc. Second Intern. Conf.Physics and Chemistry of Asbestos Minerals. 6-9. Sept. 1971., Louvain Univ., Institute of Natural Sciences, Belgium, 3-2: 9 pp.

Moody, J.B., 1976. Serpentinization: a review. Lithos, 9: 125-138.

Moore, T.R. and Zimmermann, R.C., 1977. Establishment of vegetation on serpentine asbestos mine waste, southeastern Quebec, Canada. J. Appl. Ecol., 14: 589-599.

Moore, T.R. and Zimmerman, R.C., 1979. Follow-up studies of vegetation establishment on asbestos tailings in SE-Quebec.Reclam. Review, 2: 143-146.

Morgan, A., Holmes, A. and Lally, A.E., 1971. Solubility of chrysotile asbestos and associated trace metals in N hydrochloric acid at 25 C. Proc. Second Intern. Conf. Physicsand Chemistry of Asbestos Minerals. Louvain Univ., Institute of Natural Sciences, Belgium, 2-8: 13 pp.

Morgan, A., Holmes, A. and Gold, 1971. Studies of the solubility of constituents of chrysotile asbestos in vivo using radioactive tracer techniques. Environ. Res., 4: 558-570.

Morgan, M.A., Jackson, W.A. and Volk, R.J., 1972. Nitrate absorption and assimilation in ryegrass as influenced by calcium and magnesium. Pl. Physiol., Lancaster, 50: 485-490.

Morgan, A., Lally, A.E. and Holmes, A., 1973. Some observations on the distribution of trace metals in chrysotile asbestos. Ann. Occup. Hyg., 16: 231-240.

Morgan, A., Davies, P., Wagner, J.C., Berry, G. and Holmes, A., 1977. The Biological effect of magnesium-leached chrysotile asbestos. Br. J. Exp. Path., 58:465-473.

Morgan, R.W., Foliart, D.E. and Wong, O., 1985. Asbestos and gastrointestinal cancer. Western J. of Med., 143 (1): 60-65.

Mossman, B.T. and Craighead, J.E., 1981. Mechanisms of asbestos carcinogenesis. Environ. Res. 25: 269-280.

Mueller, P.K., Alcocer, A.E., Stanley, R.L. and Smith, G.R., 1975 asbestos fiber atlas. US-Environmental Protection Agency, Chemistry and Physics Laboratory, Research Triangle Rark, N.C. EPA-650/2-75-036, 50 pp.

Mumpton, F.A. and Thompson, C.S., 1975. Mineralogy and origin of the Coalinga asbestos deposits. Clays & Clay Miner., 23: 131-143.

Ney, P.E., 1986. Asbestos. In: O. Hutzinger (Editor), The Handbook of Environmental Chemistry. Anthropogenic Compounds, Springer Verlag, N.Y., 3 (Part D): 35-100.

Nicholson, W.J., 1971. Asbestos air pollution in New York City. In: H.M. Englund and W.T. Berry (Editors), Proc. Second Clean Air Congress, N.Y., Academy Press, N.Y., pp. 136-139.

Nicholson, W.J., 1984. Environmental asbestos contamination: The situation in the U.S.A. Proc. Intern. Conf. Environmental Contamination. London, U.K., July 1984, UN Envir. Progr. (IRPTC), pp.49-56.

Nicholson, W.J., Rohl, A.N., Weisman, I. and Selikoff, I.J., 1980. Environmental asbestos concentrations in the United States. In: J.C. Wagner (Editor), Biological effects of mineral fibers. Intern. Agency for Research on Cancer, Lyon, France, Vol. 2: 823-827.

Ogura, Y., Keiichi Koide, K.I. and Shimosaka, K.,1981. Geochemistry and mineralogy of nickel oxide ores in the Southwestern Pacific area. Proc. Intern. Seminar on Lateritisation Processes, Trivandrum, India, Dec. 11-14, 1979, A.A. Balkema, Rotterdam, pp. 58-67.

Oliver, T. and Murr, L.E., 1977. An electron microscope study of asbestiform fibre concentrations in Rio Grande Valley water supplies. J. Am. Water Works Assoc., Aug.,: 428-431.

Oterdoom, W.H., 1978. Tremolite- and Diopside-bearing serpentine assemblages in the CaO-MgO-SiO2-H2O multisystem. Schweiz. Mineral. Petrogr. Mitt., 58: 127-138.

Oxberry, J.R., Doudoroff, P. and Anderson, D.W., 1978. Potential toxicity of taconite tailings to aquatic life in Lake Superior. J. Water Pollut. Control Fed., 50: 240-251.

Page, N.J. and Coleman, R.G., 1967. Serpentine-mineral analyses and physical properties. U.S. Geol. Survey Prof. Paper 575B, pp. 103-107.

Palnar, P.V., 1988. Further evidence of non-asbestos related mesothelioma. Scand. J.Work Environ., 14: 141-144.

Papirer, E., Dovergne, G. and Leroy, P., 1976. Modifications physico-chimiques du chrysotile par attaque chimique menagee. I: en milieu aqueux. Bull. Soc. Chim. Fr., 1: 651-653.

Paquet, H., Duplay, J. and Nahon, D., 1981. Variations in the composition of phyllosilicates monoparticles in a weathering profile of ultrabasic rocks. Proc. 7th Intern. Clay Conf. 1981. In: H. van Olphen and F. Veniale (Editors), Elsevier, Developments in Sedimentology, Amsterdam. 35: 595-603.

Parry, W.T., 1985. Calculated solubility of chrysotile asbestos in physiological systems. Environ. Res., 37: 410-418.

Patel-Mandlik, K.J., Manos, C.G. and Liks, D.J., 1988. Identification of asbestos and glass-fibers in sewage sludges of small New York state cities. Chemosphere, 17: 1025-1032.

Pathak, B. and Sebastien, P., 1985. Surface characterization of chrysotile asbestos by X-ray photoelectron spectroscopy and scanning auger spectroscopy. Can. Journ. of Spectroscopy, 30:1-6.

Perry, H.M., Aldon, E.F. and Brock, J.H., 1987. Reclamation of an asbestos mill waste site in southwestern United States. Reclam. Reveg. Res., 6:187-196.

Peterson, D.L., 1978. The Duluth Experience - Asbestos, water and the public. AWWA, J., 70: 24-28.

Peterson, D.L., Schleppenbach, F.S. and Zaudtke, T.M., 1980. Studies from asbestos removal by direct filtration of Lake Superior water. J. Am. Water Works Assoc. 72: 155-161.

Peterson, P.J., 1983. Adaptation to toxic metals. In: D.A. Robb and W.S. Pierpoint (Editors), Metal and micronutrients: Uptake and Utilization by Plants. Academic Press, London/New York, pp. 51-69.

Peterson, P.J. and Girling, C.A., 1983. Other trace metals. In: N.W. Lepp (Editor), Effect of heavy metal pollution on plants. Vol.1. Effects of trace metals on plant function. Applied Science Publishers, London/New Jersey, pp. 213-277.

Petrov, V.P. and Znamensky, 1981. Asbestos deposits of the USSR. In: P.H. Riordon (Editor), Geology of Asbestos Deposits, AIME, New York, pp. 45-52.

Pitt, R., 1988. Asbestos as an urban area pollutant. J. Water Poll. Contr. Fed., 11: 1993-2011.

Polissar, L., Severson,R.K., Boatman,E.S. and Thomas, D.B., 1982. Cancer Incidence in Relation to Asbestos in Drinking Water in the Puget Sound Region. Am. J. Epidemiol., 116:314-328.

Polissar, L., Severson, R.K. and Boatman, E.S., 1983. Cancer risk from asbestos in drinking water: summary of a case-control study in Western Washington. Environ. Health Perspect., 53: 57-60.

Pooley, F.D., 1976. An examination of the fibrous mineral content of asbestos lung tissue from the Canadian chrysotile mining industry. Environ. Res., 12: 281-298.

Pooley, F.D., 1986. Asbestos mineralogy. In: K. Antman and J. Aisner (Editors), Asbestos Related Malignancy. Grune and Stratton Inc., New York, San Francisco, London, pp. 3-27.

Popendorf, W. and Wenk, H.R., 1983. Chrysotile asbestos in a vehicular recreation area: A case study. In: R.H. Webb and H.G. Wilshire (Editors), Environmental Effects of Off-Road Vehicles, Springer Verlag, N.Y., pp. 375-396.

Pott, F., 1978. Some aspects on the dosimetry of the carcinogenic potency of asbestos and other fibrous dust. Staub-Reinhaltung Luft, 38: 486-490.

Pott, F, 1980. Animal experiments on biological effects of mineral fibres. In: J.C. Wagner (Editor), Biological Effects of Mineral Fibres. Intern. Agency for Research on Cancer, Vol. 2: 261-272.

Pott, F., 1987. Die Faser als krebserzeugendes Agens. Zentralblatt fur Bakt. Hyg. B., 184:1-23.

Proctor, J., 1970. Magnesium as a toxic element. Nature, 227: 742-743.

Proctor, J., 1971. The plant ecology of serpentine. II Plant response to serpentine soils. J. Ecol., 59: 397-410.

Proctor, J. and Whitten, K., 1971. A population of the valley pocket gopher (Thomomys bottae) on a serpentine soil. Am. Midl. Nat., 78: 176-179.

Proctor, J., and Woodell, S.R.J., 1971. The plant ecology of serpentine. Serpentine vegetation of England and Scotland. J. Ecol., 59: 325-395.

Proctor, J. and Woodell, S.R.J., 1975. The ecology of serpentine soils. Adv. Ecol. Res., 9: 255-366.

Proctor, J, and McGowan, I.D., 1976. Influence of magnesium on nickel toxicity. Nature, 260: 134.

Proctor, J. and Cottam, D.A., 1982. Growth of oats beet and rape in four serpentine soils. Trans. Bot. Soc. Edinb., 44: 19-25.

Puffer, J.H., Germine, M., Hurtubise, D.O., Mortec, K.A. and Bello, D.M., 1980. Asbestos distribution in the central serpentine district of Maryland-Pennsylvania. Environ. Res., 23: 233-246.

Puffer, J.H., Germine, M. and Maresca, G.P., 1987. Rutile fibers in surface waters of Northern New Jersey. Arch. Environ. Contam. Toxicol., 16: 103-109.

Rabenhorst, M.C. and Foss, J.E., 1981. Soil and geologic mapping over mafic and ultramafic parent materials in Maryland. Soil Sci. Soc. Am. J., 45: 1156-1160.

Rabenhorst, M.C., Fanning, D.S. and Foss, J.E., 1982a. Regularly interstratified chloritre/vermiculite in soils over meta-igneous mafic rocks in Maryland. Clays and Clay Minerals, 2: 156-158.

Rabenhorst, M.C., Foss, J.E. and Fanning, D.S., 1982. Genesis of Maryland soils formed from serpentinite. Soil Sci. Soc. Am. J., 46: 607-616.

Ralston, J. and Kichener,J.A., 1975. The surface chemistry of amosite asbestos, an amphibole silicate. J. Colloid and Interface Sci., 50: 242-249.

Reeves, R.D. and MacFarlane, R.M., 1981. Nickel uptake by California Streptanthus and Caulanthus with particular reference to the hyperaccumulator S. polygaloides Grey (Brassicaceae). Amer. J. Bot., 68: 708-712.

Reeves, R.D. and Baker, A.J.M., 1984. Studies on metal uptake by plants from serpentine and non-serpentine populations of Thlaspi goesingense halacsy (cruciferae). New Phytologist, 98: 191-204.

Reeves, R.D., Macfarlane, R.M. and Brooks, R.R., 1983. Accumulation of nickel and zinc by western North American genera containing serpentine-tolerant species. Amer. J. Bot., 70: 1297-1303.

Reimschussel, G. 1975. Association of trace metals with chrysotile asbestos. CIM Bull., 68: 76-83.

Rendall, R.E.G., 1980. Physical and chemical characteristics of UICC reference samples. In: J.C. Wagner (Editor), Biological Effects of Mineral Fibres. Intern. Agency for Research on Cancer. IARC Sci. Publ. No. 30: 87-95.

Rickards, A.L. and Badami, D.V. 1971. Chrysotile asbestos in urban air. Nature, 234: 93-94.

Riordon, P.H. (Editor), 1981. Geology of Asbestos Deposits. Soc. of Mining Engineers, American Institute of Mining, Metallurgical, and Petroleum Engineers Inc. New York, 111 pp.

Ritter-Studnicka, H., 1970. Die Flora der Serpentinvorkommen in Bosnien. Bibl. Bot., 130: 1-100.

Ritter-Studnicka, H. and Klement, O., 1968. On lichen species and their associations on serpentine in Bosnia. Ost. Bot. Z., 115: 93-99.

Roberts, B.A. 1980. Some chemical and physical properties of serpentine soils from Western Newfoundland. Can. J. Soil Sci., 60:231-240.

Robertson, A.I., 1985. The poisoning of roots of Zea Mays by nickel ions, and the protection afforded by magnesium and calcium. New Phytol., 100: 173-189.

Rodrique, L., 1984. Les mineraux d'amiante et leur characterisation. Revue des Questions Scientifiques, 155: 225-258.

Rohl, A.N., Langer, A.M. and Selikoff, I.R., 1977. Environmental asbestos pollution related to use of quarried serpentine rock. Science, 196: 1319-1322.

Rohl, A.N., Langer, A.M., Moncure, G., Selikoff, I.J. and Fischbein, A., 1982. Endemic pleural disease associated with exposure to mixed fibrous dust in Turkey. Science, 216: 518-520.

Ross, M., 1982. The geological occurrences of the commercial asbestos minerals and other asbestos-like minerals. Proc. World Symp. of Asbestos, Montreal, Quebec. Can. Asbestos Inf. Centre, Montreal, Quebec, Session 5, 19 pp.

Rossiter, C.E., 1987. Asbestos blues. Med. J. of Australia, 147: 162-163.

Roy-Chowdhury, A.K., Mooney, T.F. and Reeves, A.L. 1973. Trace metals in asbestos carcinogenesis. Arch. Environ. Health, 26: 253-255.

Sahu, K.C., 1981. Preliminary studies on formation of Ni-rich laterite over ultramafic rocks of Amjori sill in Similipal, Mayurbhanj District, Orissa Proc. Intern. Seminar on Lateritisation Processes, Trivandrum, India, 11-14 Dec. 1979, AA Balkema, Rotterdam, pp. 68-76.

Sasse, F., 1979. Untersuchungen an Serpentinstandorten in Frankreich, Italien, Osterreich und der Bundesrepublik Deutschland. I. Bodenanalysen. Flora, 168: 378-395.

Sasse, F., 1979, Untersuchungen an Serpentinstandorten in Frankreich, Italien, Osterreich und der Bundesrepublik Deutschland. Flora, 168: 578-594.

Schellman, W., 1964. Laterite weathering of serpentinite. Geol. Jahrb., 81: 645-678.

Schellmann, W., 1982. Formation of nickel silicate ores by weathering of ultramafic rocks. Proc. 7th Intern. Clay Conf. 1981., In: H. van Olphen and F. Veniale (Editors), Elsevier, Developments in Sedimentology, Amsterdam, 35: 623-633.

Schiller, J.P., Payne, S.L. and Khalafalla, S.E., 1980. Surface charge heterogeneity in amphibole cleavage fragments and asbestos fibers. Science, 209: 1530-1532.

Schott, J. and Berner, R.A., 1985. Dissolution mechanisms of pyroxenes and olivines during weathering. In: J.I. Drever (Editor), The Chemistry of Weathering. D. Reidel Publishing Company, pp. 35-53.

Schreier, H., 1987. Asbestos fibers introduce trace metals into streamwater and sediments. Environ. Pollut., 43: 229-242.

Schreier, H. and Taylor, J., 1980. Asbestos fibers in receiving waters. Inland waters Directorate, Environment Canada, Vancouver, B.C., Techn. Bull. No. 117 : 19 pp.

Schreier, H. and Taylor, J., 1981. Variations and Mechanisms of Asbestos Fibre Distribution in Stream Water. Inland Waters Directorate, Environment Canada, Vancouver, B.C., Techn. Bull. No. 118: 17 pp.

Schreier, H. and Timmenga, H. 1986. Earthworm response to asbestos-rich serpentinitic sediments. Soil Biol. Biochem., 1:85-89.

Schreier, H., Shelford, J.A. and Nguyen, T.D., 1986. Asbestos fibers and trace metals in the blood of cattle grazing in fields inundated by asbestos rich sediments. Environ. Res., 41: 95-109.

Schreier, H., Omueti, J.A. and Lavkulich, L.M., 1987a. Weathering processes of asbestos-rich serpentinitic sediments. Soil Sci. Soc. Am. J., 51: 993-999.

Schreier, H., Northcote, T. and Hall, K., 1987b. Trace metals in fish exposed to asbestos rich sediments. Water, Air, Soil Pollut., 35:279-291.

Schwarz, E. J. and Winer, A.A., 1971. Magnetic properties of asbestos, with special reference to the determination of absolute magnetite contents. CIM Bull., 64: 55-59.

Sebastien, P., Bignon, J., Goni, J., Dufour, G. and Bonnard, G., 1975. Electron microscopic findings in asbestos air pollution. J. Microsc., 23: 76-77.

Selikoff, I.J. and D.H.K. Lee, 1978. Asbestos and Disease. Academic Press, New York, 549 pp.

Seshan, K., 1978. On the utility of dark-field electron microscopy in the determination of the degree of deformation in chrysotile asbestos: an environmental research application. Environ. Res., 16: 383-392.

Seshan, K., 1983. How are the physical and chemical properties of chrysotile asbestos altered by a 10-year residence in water and up to 5 days in simulated stomach acid? Environ. Health Perspect., 53: 143-148.

Severne, B.C. and Brooks, R.R., 1972. A nickel accumulating plant in Western Australia. Planta, 103: 91-94.

Severson, R.K., Harvey, J. and Polissar, L., 1981. The relationship between asbestos and turbidity in raw water. Journ. AWWA., 73: 223-224.

148

Shacklette, H.T. and Boerngen, J.G., 1984. Element concentrations in soils and other surficial materials of the conterminous United States. US Geological Survey, Prof. Paper, 1270: 105 pp.

Sheils, A.K., 1984. The disposal of asbestos wastes in the U.K. Proc. Intern. Conf. Environmental Contamination. London, U.K. July 1984, UN Envir. Progr. (IRPTC), pp. 57-62.

Shewry, P.R. and Peterson, P.J., 1974. The uptake and transport of chromium by barley seedlings (Hordeum vulgare). J. Exp. Bot., 25: 785-97.

Shewry, P.R. and Peterson, P.J., 1975. Calcium and magnesium in plants and soils from a serpentine area on Unst., Shetland. J. Appl. Ecol., 12: 381-391.

Shewry, P.R. and Peterson, P.J., 1976. Distribution of chromium and nickel in plants and soil from serpentine and other sites. J. Ecology, 64: 195-212.

Shugar, S., 1979. Effects of asbestos in the Canadian Environment. National Research Council Canada, Assoc. Committee on Sci. Criteria for Environ. Quality, NRCC, No. 16452: 185 pp.

Siegrist, H.G. and Wylie, A.G., 1980. Characterizing and discriminating the shape of asbestos particles. Environ. Res., 23: 348-361.

Sigurdson, E.E., 1983. Observations of cancer incidence in Duluth, Minnesota.Environ. Health Perspect. 53: 61-67.

Sigurdson, E.E., Levy, B.S., Mandel, J., McHugh, R., Michienzi, L.J., Jagger, H. and Pearson, J., 1981. Cancer morbidity investigations: lessons for the Duluth study of possible effects of asbestos in drinking water. Environ. Res., 25: 50-61.

Skippen, G. and McKinstry, B.W., 1985. Synthetic and natural tremolite in equilibrium with fosterite, enstatite, diopside and fluid. Contributions to Mineralogy and Petogeology, 89: 256-262.

Slingsby D.R. and Brown, D.H., 1977. Nickel in British serpentine soils. Journ. Ecology. 65:597-618.

Smith, R.A.H. and Bradshaw, A.D., 1979. The use of metal tolerant plant populations for the reclamation of metalliferous wastes. J. Appl. Ecol., 16: 595-612.

Smith, W.E., Hubert, D.D., Sobel, H.J., Peters, E.T. and Doerfler, T.E., 1980. Health of experimental animals drinking water with and without amosite asbestos and other mineral particles. J. Env. Path. Tox., 3: 277-330.

Soane, B.D. and Saunder, D.H., 1959. Nickel and chromium toxicity of serpentine soils in Southern Rhodesia. Soil Sci., 88: 322-330.

Spiel, S. and Leineweber, J.P., 1969. Asbestos minerals in modern technology. Environ. Res., 2: 166-201.

Spurny, K.R., 1981. Mechano-chemical and physio-chemical changes of some mineral fibrous particles under environmental and biological conditions. In: K. Iinoya, J.K. Beddow, and G. Jimbo (Editors), Powder Technology. Hemisphere Publ. Washington, pp. 427-434.

Spurny, K.R., 1982. On the problem of measuring and analysis of chemically changed mineral fibers in the environment and in biological materials. The Sci. of the Total Environ., 23: 239-249.

Spurny, K.R., 1983. Measurement and analysis of chemically changed mineral fibers after experiments in vitro and in vivo. Env. Health Persp., 51: 343-355.

Spurney,.K.R., 1986. On the filtration of fibrous aerosols. The Sci. of the Total Environ., 52: 189-199.

Spurny, K.R., Pott, F., Huth, F., Weiss, G. and Opiela, H., 1979. Identizierung und krebserzeugende Wirkung von faserformigen Aktinolith aus einem Steinbruch. Staub-Reinhalt. Luft, 39: 386-389.

Spurny, K.R. and Schormann, T., 1983. Faserformige Partikeln und Wasseranalytik: einige vorlaufige Erbergebnisse der Trinkwasseranalysen in der Bundersrepublik Deutschland. Z. Wass. Abwass. Forsch. 16: 24-26.

Stanton, M.F. and Layard, M., 1978. The carcinogenicity of fibrous minerals. In: C.C. Gravatt, P.D. LaFleur, and K.F.J. Heinreich (Editors), Workshop on Asbestos: Definition and Measurement Methods (National Bureau of Standards Special Publication No. 506, Wash. D.C., National Measurement Laboratory, pp. 143-151.

Stanton, M.F., Layard, M., Tegeris, A., Miller, E., May, M. and Graf, E., 1977. Relationship of particle dimensions to the carcinogenicity in amphibole asbestos and other fibrous minerals. J. Natl. Cancer Inst. 58: 587-603.

Stanton, M.F., Layard, M., Tegeris, A., Miller, E., May, M., Morgan, E. and Smith, A., 1981. Relation of particle dimension to carcinogenicity in amphibole asbestosis and other fibrous minerals. J. Natl. Cancer Inst., 67: 965-975.

Stebbins, G.L., 1942. The genetic approach to rare and endemic species. Madrono, 6: 241-272.

Steel, E. and Wylie, A., 1981. Mineralogical characteristics of asbestos. In: P.H. Riordon (Editor), Geology of Asbestos Deposits. Soc. of Mining Engineers, AIME, N.Y., pp. 93-100.

Steinbauer, J., Boutin, C., Viallat, J.R., Dufour, G., Gaudichet, A., Massey, D.G., Charpin, D. and Mouries, J.C., 1987. Plaques pleurales et environement asbestosique en Corse du Nord. Revue des maladies respiratoires, 4 (1): 23-27.

Stewart, H.L., 1977. Discussion Paper: Enigmas of cancer in relation to neoplasms of aquatic animals. Ann. M.Y. Acad. Sci., 298: 305-315.

Stewart, I., 1976. Asbestos in the water supply of ten regional cities. W.C. McCrone Associates Inc., EPA-560/6-76-017, 58 pp.

Stewart, I., 1978. Transmission electron microscopical methods for the determination of asbestos. Proc. of Workshop on Asbestos: Definitions and Measurement Methods. N.B.S. Special Publ. 506: 271-280.

Stewart, I., Putscher, R.E., Humecki, H.J. and Shrimp, R.J., 1976. Asbestos fibers in natural runoff on discharge from sources manufacturing asbestos products. W.C. McCrone Assoc. Inc., for EPA-560/6-76-020.

Stewart, R.V., 1981. Geology and evaluation of the asbestos hill orebody. In: P.H. Riordon (Editor), Geology of Asbestos Deposits, AIME, New York, pp. 53-62.

Sticher, H., Bach, R., Brugger, H. and Voekt, U., 1975. Flugstaub in vier Boden aus Kalk, Dolomit und Serpentine. Catena, 2: 11-22.

Sticher, H., Gasser, U. and Juchler, S., 1986. Die Boden auf Serpentinit bei Davos: Entstehung, Verbreitung, Eigenschaften. Veroff. Geobot. Inst. ETH, Zurich, 87: 275-290.

Stroink, G., Hutt, D., Lim, D. and Dunlap, R.A., 1985. The magnetic properties of chrysotile asbestos. IEEE Trans. Magnetics, 21: 2074-2076.

Subbanna, G.N., Kutty, T.R.N. and Anantha-Iyer, G.V., 1986. Structural intergrowth of brucite in anthophyllite. Am. Mineralogist, 71: 1198-1200.

Suquet, H., Malard, C., Fournier, J. and Pezerat, H., 1987. Capacite d'echange cationique et charge de surface du chrysotile. Bull. Mineral., 110: 711-715.

Sullivan, J., 1986. American Water Works Association opposes ban on AC-pipes. AWWA J., 78, p.16.

Swaine, D.J. and Mitchell, R.J., 1960. Trace element distribution in soil profiles. J. Soil Sci., 11: 347-368.

Swenters, I.M., De Waele, J.K., Verlinden, J.A. and Adams, F.C., 1985. Comparison of secondary-ion mass spectrometry and compleximetric titration for the determination of leaching of magnesium from chrysotile asbestos. Analytica Chimica Acta, 173: 377-380.

Szilagyi, M., 1967. Sorption of molybdenum by humus preparations. Geochem. Intern., 4: 1165-1167.

Taylor, G.J. and Crowder, A.A., 1983. Uptake and accumulation of copper, nickel, and iron by Typha latifolia grown in solution culture. Can. J. Bot., 61: 1825-1830.

Tartar, E.M., Cooper, R.C. and Freeman, W.R., 1983. A Graphical analysis of the interrelationships amongst waterborne asbestos digestive system cancer and population density. Environ. Health Perspect., 53: 88-89.

Teherani, D.K., 1985. Determination of arsenic, scandium, chromium, cobalt and nickel in asbestos by neutron activation analysis. J. Radional. Nucl. Chem., 95: 177-184.

Thomassin, J.H., Goni, J., Baillif, P., Touray, J.C. and Jaurand, M.C., 1977. An XPS study of the dissolution kinetics of chrysotile in 0.1 N Oxalic acid at different temperatures. Phys. Chem. Minerals, 1: 385-398.

Thomassin, J.H., Touray, J.C., Bailif, P., Jaurand, M.C., Magne, L. and Goni, J., 1980. Surface interaction between chrysotile and solutions (dissolution and adsorption): Systematic X-ray photoelectron spectroscopy studies. In: J.C. Wagner (Editor), Biological Effects of Mineral Fibres. Intern. Agency for Research on Cancer. IARC Sci. Publ. No. 30: 105-112.

Thompson, M.R., 1976. Procedure for examination of water and sediment samples for total asbestos fibre count by electron microscopy. Environment Canada, Inland Waters Directorate, Burlington, Ontario, Techn. Bull. No. 94, 4 pp.

Thompson, R.J., 1978. Ambient air monitoring for chrysotile in the United States. Workshop on asbestos: Definitions of measurement methods. N.B.S. Spec. Publ. 506: 355-362.

Thornton, I., 1981. Geochemical aspects of the distribution and forms of heavy metals in soils. In: N.W. Lepp (Editor), Effect of heavy metal pollution on plants. Applied Science Publ., London/New Jersey, Vol. 2: pp. 1-33.

Timbrell, V., 1969. Characteristics of the International Union Against Cancer standard reference samples of asbestos. Pneumoconiosis, Proc. Intern. Conf., Johannesburg, Oxford Univ. Press, Cape Town, S.A. pp. 28-36.

Timbrell, V., 1975. Alignment of respirable asbestos fibers by magnetic fields. Annals Occup. Hyg., 18: 299-311.

Timbrell, V., Gilson, J.C. and Webster, I., 1968. UICC standard reference samples of asbestos. Int. Journ. Cancer, 3: 406-408.

Timbrell, V. Pooley, F.D. and Wagner, J.C., 1970. Characteristics of respirable asbestos fibers. In: H.A. Shapiro (Editor), Pneumoconiosis: Proceedings of the Intern. Conf., Johannesburg, 1969, Cape Town, Oxford Univ. Press, Oxford, pp. 120-125.

Timbrell, V, Griffiths, D.M. and Pooley, F.D., 1971. Possible biological importance of fibre diameters of South African amphiboles. Nature, 232: 55-56.

Timperley, M.H., Brooks, R.R. and Peterson, P.J., 1970. The significance of essential and non-essential trace elements in plants in relation to biogeochemical prospecting. J. Appl. Ecol., 7: 429-439.

Toft, P. and Meek, M.E., 1986. Human exposure to asbestos in the environment. In: J.N. Lester (Editor), Proc. Intern. Conf. Chemicals in the Environment, July 1-3, 1986, Lisbon, Portugal, Selper Ltd. London, U.K. pp. 492-500.

Toft, P., Wigle, D., Meranger, J.C. and Mao, Y., 1981. Asbestos and drinking water in Canada. Sci. Total Environ., 18: 77-89.

Toft, P., Meek, M.E., Wigle, D.T. and Meranger, J.C., 1984. Asbestos in drinking water. CRC Critical Reviews in Environmental Control, 14: 151-197.

Trescases, J.J., 1975. L'evolution geochimique supergene des rockes ultrabasiques en zone tropicale. Mem. O.R.S.T.O.M., 78: 1-259.

Upreti, R.K., Dogra, R.K.S., Shanker, R., Krishna Murti, C.R., Dwivedi, K.K. and Rao, G.N., 1984. Trace elemental analysis of asbestos with an X-ray fluorescence technique. The Sci. of the Total Environ., 40: 259-267.

Ure, A.M. and Berrow, M.L., 1982. The elemental constituents of soils. Environm. Chemistry, 2: 94-204.

Veblen, D.R., Buseck, P.R. and Burnham, C.W., 1977. Asbestiform chain silicates: New minerals and structural groups. Science, 198: 359-365.

Velema, J.P., 1987. Contaminated drinking water as a potential cause of cancer in humans. Envir. Carcino. Rev., J. Envir. Sci. Health., C5: 1-28.

Verlinden, J.A., De Waele, J.K., Swenters, I.M. and Adams, F.C., 1984. Secondary ion mass spectrometric analysis of leaching behaviour of magnesium from chrysotile in oxalic acid solution. Surf. and Interface Analysis, 6: 286-290.

Verlinden, J.A., De Waele, J.K., Swenters, L.M. and Adams, F.C., 1985. Surface adsorption behaviour of amosite asbestos fibers as studied by SIMS and LAMMA. Spectrochimica Acta, 40B: 859-864.

Virta, R.L. and Segreti, J.M., 1987. A model for predicting crocidolite fiber size distributions. Environ. Res., 44: 148-160.

Wagner, J.C. (Editor), 1980. Biological Effects of Mineral Fibers. IARC Monographs on Evaluation of Carcinogenic Risksof Chemicals to Man, IARC Sci. Publ. No. 30, Vol 1 & 2, 1007 pp.

Wagner, J.C., 1986. Mesothelioma and mineral fibers. Cancer, 57: 1905-1911.

Wagner, J.C. and Berry, G., 1969. Mesotheliomas in rats following inoculation with asbestos. Br. J. Cancer, 23: 567.

Wagner, J.C. and Pooley, F.D., 1986. Mineral fibres and mesothelioma. Thorax, 41: 161-165.

Wagner, J.C., Sleggs, C.A. and Marchand, P., 1960. Diffuse pleural mesothelioma and asbestos exposure in Northwestern Cape Province. Br. J. Ind. Med. 17: 260-271.

Wagner, J.C., Berry, G. and Timbrell, V., 1973. Mesotheliomata in rats after inoculation with asbestos and other minerals. Br.J. Cancer., 28: 173-185.

Wagner, J.C., Berry, G., Cooke, T.J., Hill, R.J., Pooley, F.D. and Skidmore, J.W., 1975. Inhaled Particles. Pergamon Press, 647 pp.

Wagner, J.C., Skidmore, J.W., Hill, R.J. and Griffiths, D.M., 1985. Erionite exposure and mesotheliomas in rats. Br. J. Cancer, 51: 727-730.

Wagner, J.C., Newhouse, M.L., Corrin, B., Rossiter, C.E.R. and Griffith, D.M., 1988. Correlation between fiber content of the lung and disease in East London asbestos factory workers. Brit. J. Ind. Med., 45: 305-308.

Walker, R.B., Walker, H.M. and Ashworth, P.R., 1955. Calcium-magnesium nutrition with special reference to serpentine soils. Pl. Physiol., 30: 214-221.

152

Wallace, A., Jones, M.B. and Alexander, G.V., 1982. Mineral composition of native woody plants growing on serpentine soil in California. Soil Sci., 134: 42-44.

Webber, J.S., S. Syrotynski, and M.V. King., 1988. Asbestos-Contaminated Drinking Water: Its Impact on Household Air. Environm. Res., 46: 153-167.

Webster, I., 1974. The ingestion of asbestos fibers. Environ. Health Perspect., 9: 199-202

Weiss, W., 1986. History of hazards associated with asbestos, Pennsylvania Medicine, 89: 57-60.

White, C.D., 1967. Absence of nodule formation on Ceanothus cruneatu in serpentine soils. Nature, 215: 875 pp.

Whittaker, E.J.W., 1979. Mineralogy, chemistry and chrystallography of amphibole asbestos. In: R.L. Ledoux (Editor), Short Course in Mineralogical Techniques of Asbestos Determination. Mineralogical Association of Canada, 4 (Section 1, Part A): 1-32.

Whittaker, E.J.W. and Wicks, F.J., 1970. Chemical differences among the serpentine "polymorphs": A discussion. Am. Mineral., 55: 1025-1047.

Whittaker, E.J.W. and Middleton, A.P., 1979. The intergrowth of fibrous brucite and fibrous magnesite with chrysotile. Canadian Mineralogist, 17: 699-702.

Whittaker, R.H., 1954. The vegetational response to serpentine soils. In: The ecology of serpentine soils: A symposium. Ecology, 35: 275-288.

Wicks, F.J., 1979. Mineralogy, Chemistry and chrystallography of chrysotile asbestos. In: R.L. Ledoux (Editor), Short Course in Mineralogical Techniques of Asbestos Determination. Mineralogical Association of Canada, 4 (Section 1, Part B):   35-78.

Wicks, F.J. and Whittaker, E.J.W., 1975. A reappraisal of the structures of the serpentine minerals. Canadian Mineralogist, 13: 227-243.

Wicks, F.J. and Whittaker, E.J.W., 1977. Serpentine textures and serpentinization. Canadian Mineralogist, 15: 459-488.

Wigle, D.T., 1977. Cancer Mortality in Relation to Asbestos in the Municipal Water Supplies. Arch. Environ. Health, 32:185-190.

Wigle, D.T., Mao, Y., Semenciw, R., Smith, M.H. and Toft, P., 1986. Contaminants in drinking water and cancer risks in Canadian cities. Can. J. Public Health, 77: 335-342.

Wild, H., 1970. Geobotanical anomalies in Rhodesia 3.- The vegetation of nickel-bearing soils. Kirkia, Suppl. Vol.7: 1-62.

Wild, H., 1975. Termites and serpentines of the Great Dyke of Rhodesia. Trans. Rhodesia Scientific Assoc., 57: 1-11.

Wildman, W.E., Jackson, M.L. and Whittig, L.D. 1968a. Iron rich montmorillonite formation in soils derived from serpentinite. Soil Sci. Soc. Am. J., 32:787-794.

Wildman, W.E., Jackson, M.L. and Whittig, L.D., 1968b. Serpentine rock dissolution as a function of carbon dioxide pressure in aqueous solution. Am. Miner. 53: 1252-1263.

Wildman, W.E., Whittig, L.D. and Jackson, M.L., 1971. Serpentine stability in relation to the formation of iron-rich montmorillonite in some Californian soils. Am. Mineralogist, 56:587-602.

Willey, R.J., 1987. Magnetic orientation of respirable asbestos fibers. J. Applied Physics, 61: 3214-3261.

Williams, H., Hibbard, J.P. and Bursnall, J.T., 1977. Geological setting of asbestos-bearing ultramafic rocks along the Baie Verte Lineament, Newfoundland, Report on Activities, Part A; Geol. Surv. Canada Paper 77-1A, pp. 351-360.

Wilson, M.J. and Berrow, M.L., 1978. The mineralogy and heavy metal content of some serpentinitic soils in SE Scotland. Chem. Erde, 37: 181-205.

Wilson, M.J., Jones, D. and McHardy N.J., 1981. The weathering of serpentine by <u>Lecanora atra.</u> Lichenologist, 13: 167-176.

Wither, E.D. and Brooks, R.R., 1977. Hyperaccumulation of nickel by some plants of southeast Asia. J. Geochem. Expl., 8: 579-583.

Wolf, K.M., Piotrowski, Z,.H., Engel, J.D., Bekeris, L.G., Palacios, E. and Fisher, K.A., 1987. Malignant mesothelioma with occupational and environmental asbestos exposure in an Illinois Community Hospital. Archives of Internal Medicine, 147 (12): 2145-2149.

Wylie, A.G., Virta, R.L. and Segreti, J.M., 1987. Characterization of mineral population by index particle: Implication for the Stanton hypothesis. Environ. Res. 43: 427-439.

Wyllie, P.J., 1967. Ultramafic and ultrabasic rocks: petrography and petrology. In: P.J. Wyllie (Editor), Ultramafic and related rocks. John Wiley & Sons, New York, 464 pp.

Yada, K., 1967. Study of chrysotile asbestos by a high resolution electron microscope. Acta Cryst., 23: 704-707.

Yada, K., 1971. Study of microstructures of chrysotile asbestos by high resolution electron microscopy. Acta Cryst., A27: 659-664.

Yada, K. and Iishi, K., 1974. Serpentine minerals hydrothermally synthesized and their microstructures. J. Crystal Growth, 24/25: 627-630.

Yang, X.H., Brooks, R.R., Jaffre, T, and Lee, J., 1985. Elemental levels and relationships in the <u>Flacourtiaceae</u> of New Coledonia and their significance for the evaluation of the serpentine

Yazicioglu, S., 1976. Pleural calcification associated with exposure to chrysotile asbestos in southeast Turkey. Chest, 70: 43-47.

Yazicioglu, S., Ilcayto, R., Balci, B.S., Sayli, B.S. and Yorulmaz, B., 1980. Pleural calcification, pleural mesothelioma, and bronchial cancer caused by tremolite dust. Thorax, 35: 564-569.

Zoltai, T., 1978. History of asbestos-related mineralogical terminology. Proc. of Workshop an Asbestos: Definitions and Measurement Methods. N.B.S. Special Publ. 506 : 1-18.

Zussman, J., 1979. The mineralogy of asbestos. In: L. Michaels and S.S. Chissick (Editors), Asbestos; Properties, Applications, and Hazards. John Wiley and Sons, N.Y., 1 45-65.

154

SUBJECT INDEX